FLORA ZAMBESIACA

Flora terrarum Zambesii aquis conjunctarum

VOLUME TWELVE: PART ONE

FLORA ZAMBESIACA

MOZAMBIQUE

MALAWI, ZAMBIA, ZIMBABWE

BOTSWANA

VOLUME TWELVE: PART ONE

Edited by
J.R. TIMBERLAKE & E.S. MARTINS

on behalf of the Editorial Board:

D.J. MABBERLEY
Royal Botanic Gardens, Kew

M.A. DINIZ
*Centro de Botânica, Instituto de Investigação
Científica Tropical, Lisboa*

J.R. TIMBERLAKE
Royal Botanic Gardens, Kew

Published by the Royal Botanic Gardens, Kew
for the Flora Zambesiaca Managing Committee
2012

ROYAL BOTANIC GARDENS

First published in 2012 by
Royal Botanic Gardens, Kew,
Richmond, Surrey, TW9 3AB, UK
www.kew.org

ISBN 978-1-84246-374-1

British Library Cataloguing in Publication Data
A catalogue record for this book is available from the British Library

Typesetting by Christine Beard
Publishing, Design and Photography
Royal Botanic Gardens, Kew

Printed in the USA by The University of Chicago Press

For information or to purchase all Kew titles please visit
www.kewbooks.com or email publishing@kew.org

Kew's mission is to inspire and deliver science-based plant conservation worldwide,
enhancing the quality of life.

Kew receives half of its running costs from Government through the Department
for Environment, Food and Rural Affairs (Defra). All other funding needed to
support Kew's vital work comes from members, foundations, donors and
commercial activities including book sales.

CONTENTS

FAMILIES INCLUDED IN VOLUME 12, PART 1

ARACEAE (including LEMNACEAE)

LIST OF NEW NAMES PUBLISHED IN THIS PART

ARACEAE

by A. Haigh and P.C. Boyce[1]

Gigantic to minute herbs, perennial, terrestrial, epiphytic or aquatic (free floating or rooted), often with milky, viscid or acrid sap; roots often aerial, hair-like or absent in the Lemnoideae. Stems lianescent, tuberous, rhizomatous, or not differentiated into stem or leaf (Lemnoideae), in which case the plant is reduced to a minute, fleshy or flattened plant body bearing hair-like roots on undersurface, or roots absent. Cataphylls often variously ribbed and persistent, may remain intact or weather into fibres. Leaves alternate, sometimes distichous, 1 to many; normally differentiated into petiole and expanded blade; blade and petiole often variegated or mottled with various shades of green, yellow and grey. Petioles often elongate, usually smooth, rarely prickly (*Anchomanes*), typically with a distinct basal sheath, often pulvinate at or near apex, rarely centrally (*Gonatopus boivinii*). Leaf blades simple to compound, very variable in size and shape, from elliptic or ovate to sagittate or hastate, less commonly trifid to trisect, dracontioid (i.e. trisect with each primary division further much divided), pinnatifid or pinnatisect to quadripinnatifid (*Gonatopus*); sometimes perforated; midrib almost always differentiated, primary veins usually pinnate, palmate, rarely parallel, finer venation reticulate or striate. Inflorescences terminal or axillary, solitary or clustered in axils, an unbranched spadix (spike) subtended by a single spathe (bract), in Lemnoideae the inflorescence is within a minute dorsal cavity of the plant body (*Wolffia*, *Wolffiella*) or in paired lateral budding pouches (*Spirodela*, *Landoltia*, *Lemna*); spathe herbaceous, free or adnate to spadix, spreading, reflexed or convolute, sometimes constricted below middle and differentiated into tube below and blade above, often persistent, blade sometimes deciduous, green, or blade and tube of spathe differently coloured; spadix usually cylindric, erect, often fleshy and thick, flowers dense, often divided into distinct floral zones with lower part often female and a male zone above, sterile flowers of varying shape often present at base, middle or apex; apical portion sometimes forming a sterile appendix. Flowers usually numerous, minute, sessile, bractless, 2–3-merous, bisexual or unisexual, protogynous, naked or with a perigone (perianth); perigone usually thickened, cup-like or composed of 4–6 free or ± united tepals. Androecium of (1)3–6(9) stamens, free or united into synandria; anthers sessile or with elongated filaments, opening by lateral or apical slits or pores; connective often very thick. Gynoecium syncarpous; ovary normally superior, 1 to many-locular; locules each with 1 to many ovules; placentas parietal, axile, basal or apical; stigma sessile or borne on short, conical, rarely attenuate style. Fruit a 1 to many-seeded berry, or utricles (Lemnoideae). Seeds minute to large, variable in shape, with or without endosperm.

A ± cosmopolitan family with c.116 genera and over 3400 species, most abundant and diverse in humid tropical regions.

The monograph by Mayo, Bogner & Boyce (Gen. Araceae, 1997) provides a thorough coverage of the family. However, under the APG system the Araceae now includes the duckweeds in the subfamily Lemnoideae (formerly in a separate family, Lemnaceae). The CATE Araceae website (www.cate-araceae.org) has illustrated interactive keys and descriptions for all taxa in the family.

Aroids are widely used as house and garden plants across the world, especially in the tropics. As one of the most important groups in tropical horticulture, many hundreds of species may be found in cultivation. The following introduced species are included here either because they are commonly cultivated or naturalised in the Flora area, or based on a reliable record. They have not been included in the generic key.

---

[1] Former Lemnaceae written by Peter Boyce.

Alocasia odora (Lindl.) K. Koch   Elephant's Ear

Perennial terrestrial herb to 2.5 m, latex slightly milky. Stem erect to decumbent. Leaves several, clustered at tips of stems; petiole to 1.5 m; leaf blade peltate, cordate-sagittate or cordate-ovate, up to 1.3 m long. Inflorescences 2–3 together among leaf bases. Spathe up to 25 cm long, constricted ¹/₆ of way from base. Flowers unisexual, perigone absent. Fruits scarlet.

Zimbabwe. C: Harare garden (cultivated), 11.ix.1970 *Biegel* 3398 (SRGH).

Native to China and SE Asia; possibly cultivated in the Flora area. Also recorded for Zimbabwe in Maroyi (Kirkia **18**: 185, 2006), but specimen not seen so true identity not confirmed; it could be a misidentified *A. macrorrhizos* which is more commonly cultivated throughout the tropics.

Alocasia odora differs from *A. macrorrhizos* in having shortly peltate leaves vs. leaves deeply cordate at base with lateral lobes partially naked in the sinus in *A. macrorrhizos*. The most similar species in the Flora area are *Remusatia vivipara* and *Colocasia esculenta*, both of which have strongly peltate leaves.

Monstera deliciosa Liebm.   Swiss Cheese Plant

Climbing or sprawling plant with stout green internodes. Leaf-blade broadly ovate, up to 90 × 75 cm, regularly pinnatifid, adult leaves perforated in several parallel series, cordate to subcordate; petiole longer than blade, up to 1m long, pulvinate apically, sheath with persistent margins, reaching to petiole-apex. Flowers extremely infrequently; spathe broadly boat-shaped, soon deciduous; spadix cylindric. Flowers bisexual, perigone absent. (Description taken from F.T.E.A.)

Recorded from Zimbabwe in the Flora area (Biegel, Gard. Ornam. Pl. Rhod., 1977) and from Harare gardens (M. Hyde, pers. comm. 2011), but no specimens seen.

A common ornamental climber throughout the world. The ripe fruits are edible although the outer 'rind' of thickened styles should be avoided, being full of needle-like trichosclereids.

Colocasia esculenta (L.) Schott   Elephant's Ear, Madumbi

Robust herb; stem a swollen starchy tuber, sometimes forming stolons. Leaves several, petiole ± erect, sheathing for ¹/₃–²/₃ their length, blade broadly ovate in outline, peltate, cordate-sagittate, held pendent from petiole, adaxially matt waxy-glaucous and water-shedding; apex obtuse, basal lobes ± rounded; primary lateral veins 6–10. Flowering very rare in some cultivars. Spathe 20–30 cm long, basal tube green, short, limb yellow to orange, much longer; spadix shorter than spathe with conic sterile appendix to 4 cm long, occasionally reduced or absent. Flowers unisexual, perigone absent.

Zambia. W: Solwezi Dist., x.1934, *Trapnell* 1634 (K). **Zimbabwe.** E: Nyanga Dist., Stapleford, Nyamkwarara R., 3.iv.1962, *Wild* 5683 (K). Chimanimani Dist., Rusitu R., 12.i.1969, *Mavi* 900 (K). **Malawi.** S: Zomba Dist., Zomba, 9.v.1980, *Brummitt* 15612 (K).

Probably originated in Asia but has been cultivated by man since ancient times. Naturalized in Zimbabwe, Malawi, Zambia, and probably elsewhere in the Flora area, although no other specimens seen. Naturalized on river banks.

Cultivated throughout the tropics for its tubers, petioles and young leaves, which are eaten after cooking.

Xanthosoma sagittifolium (L.) Schott   Elephant's Ear (*Xanthosoma mafaffa* Schott, *Xanthosoma violaceum* Schott)

Large fleshy herb with underground tuber, and a short, stout, aerial stem in large specimens. Cut tissues exude a milky sap. Leaves forming a rosette; blade sagittate, up to c.70 × 50 cm, deep green above, paler and pruinose below with subacute basal lobes, 100 cm long or more. Spathe to 30 cm long; tube green, inflated, persistent, to 8 cm long; blade longer, creamy white to yellowish-buff, erect, deciduous; spadix slightly shorter than spathe, lacking sterile appendix. Flowers unisexual, perigone absent. (Description taken from F.T.E.A.)

Zimbabwe. E: Chimanimani Dist., Makurupini Forest, 400 m, st. 4.xii.1964, *Wild, Goldsmith & Müller* 6663 (K).
Native to tropical America. Widely cultivated in Africa for the edible, starchy tuber and leaves. Possibly naturalized in places.

Key to genera

1. Floating aquatic 2
 – Terrestrial plant, rooted aquatic or lianescent epiphyte. 7
2. Tiny plant, frond less than 1 cm long, with or without roots (unbranched if present), glabrous 3
 – Larger plant, frond 3 cm or more long, roots much branched, densely pubescent. . . . . **13. Pistia**
3. Roots present; frond with veins 4
 – Roots lacking; frond without veins 6
4. Root only 1 per frond; veins 1–5 (rarely 7 in *Lemna gibba*); base of frond without scale . . . . **2. Lemna**
 – Roots 2–21; veins usually 5–16; frond surrounded at base by a small scale 5
5. Roots usually 7–21, of which only 1–5 (never all) perforate the scale; veins usually 7–16; leaf surrounding the flowers with narrow opening at top . . . . **1. Spirodela**
 – Roots usually 2–7, all of which perforate the scale; veins usually 5–7; leaf surrounding the flowers open on one side . . . . **3. Landoltia**
6. Frond flat, with air spaces, daughter frond produced in a terminal flat pouch . . . .**4. Wolffiella**
 – Frond globose to ovoid, boat-shaped, without airspaces, daughter frond produced in a terminal conical cavity of mother frond . . . .**5. Wolffia**
7. Flowers with obvious perianth of free and fused tepals 8
 – Flowers without perianth of free or fused tepals. 10
8. Leaf pinnatisect to tri- or quadripinnatifid; tepals free; spathe margins free . . .9
 – Leaf entire, linear to cordate, sagittate or hastate; tepals united into a cup; spathe margins united at base . . . . **6. Stylochaeton**
9. Leaf blade pinnatisect; stamens free . . . . **7. Zamioculcas**
 – Leaf blade bipinnatifid to quadripinnatifid, at least in lowest pinnae; stamen filaments united. . . . .**8. Gonatopus**
10. Higher order leaf venation parallel-pinnate . . . .**11. Zantedeschia**
 – Higher order leaf venation reticulate. 11
11. Leaf blade dracontioid (trisect with each primary division further much divided), solitary leaf in each growth period. 12
 – Leaf blade of various types but never dracontioid; usually several leaves present 13
12. Petiole usually prickly; at least some of ultimate leaf lobes truncate and trapezoid; veins not forming regular submarginal collective veins on each side . . . . **9. Anchomanes**
 – Petiole usually smooth, sometimes rugose but never prickly; ultimate leaf-lobes usually oblong-elliptic, acuminate; primary lateral veins forming regular submarginal collective veins on each side . . . . **12. Amorphophallus**
13. Lianescent, root-climbing epiphyte, rarely a ground creeper; stem elongate, aerial. . . . . **10. Culcasia**
 – Terrestrial or epiphytic herb, stem abreviated, never lianescent 14
14. Leaf blade deeply pedatifid to pedatisect; never peltate. . . . . **15. Sauromatum**
 – Leaf-blade cordate-sagittate; peltate. . . . . **14. Remusatia**

ARACEAE

1. **SPIRODELA** Schleid.

Spirodela Schleid. in Linnaea **13**: 391 (1839). —Daubs, Monogr. Lemnac.: 8–16 (1965). —Hartog & Plas in Blumea **18**: 358–360 (1970). —Landolt, Biosyst. Invest. Fam. Duckweeds, Family Lemnaceae **1**: 464–471 (1986).

Fronds lanceolate, ovate or suborbicular, floating on water surface, 1 to several together, sometimes forming rosettes, surrounded basally by a bifid scale-like prophyll; frond with 5–21 'veins' and two types of crystal cells (raphides and druses). Roots 2–21 per frond. Staminate flowers 2; anther bilocular with external theca situated at same level or slightly higher than internal theca, dehiscing transversely; pistil with 1 amphitropous or 2–5 anatropous ovules. Fruit a dry pericarp with 1–2 ribbed seeds.

A genus of 2 species distributed worldwide, with one species in the Flora area.

Spirodela polyrrhiza (L.) Schleid. in Linnaea **13**: 392 (1839). —Hepper in F.W.T.A., ed.2 **3**: 129 (1968). —Hepper in F.T.E.A., Lemnaceae: 5 (1973). —Gibbs Russell in Kirkia **10**: 462 (1977). —Cook, Aq. Wetl. Pl. Sthn. Africa: 159 (2004). Type: Not known, probably from Europe (see Landolt 1986: 440). FIGURE 12.1.1.1.
    Lemna polyrrhiza L., Sp. Pl.: 970 (1753) (*polyrhiza*, see Landolt 1986: 440).

Frond ± suborbicular, 1–1.5 times long as wide, 1.5–10 × 1.5–8 mm. Roots 7–21. Inflorescence rarely produced. Pistil with 1 (rarely 2) amphitropous ovules, style c.0.3 mm long. Fruit 1–1.5 mm in diameter, with 1–2 seeds. Seeds 1–2, 0.7–1 × 0.7 mm, with 12–20 ribs.

Botswana. N: Okavango Delta, Duba (Xesabe) Is., Okavango R., 28.vi.1974, *P.A. Smith* 1061 [mixed with *Wolffiella welwitschii*] (K, SRGH). **Zambia**. C: Mpika Dist., S Luangwa Nat. Park, Mfuwe Lagoon, 9.ii.1970, *Astle* 5777 (K). E: Chipata Dist., Jumbe, Luangwa Valley, 26.xi.1966, *Mutimushi* 1675 (K, NGO, SRGH). S: Mazabuka Dist., Mazabuka, Kafue R., 3.ix.1947, *Brenan* s.n. (K spirit coll. 24318). **Zimbabwe**. N: Kariba Dist., Kariba Gorge, ix.1960, *Goldsmith* 106a/60 [mixed with *Lemna aequinoctialis* as *Goldsmith* 106b/60] (SRGH). C: Kadoma Dist., Ngezi, n.d., *Konietzko* 36 (BR). **Malawi**. N: Mzimba Dist., Lake Kazuni, n.d., *Pawek* 3354 (K). S: Blantyre Dist., Kogodzi R. on road to Nkula Falls, 9 km NW of Chileka, 630 m, 2.vii.1970, *Brummitt* 11780 (K, SRGH). **Mozambique**. Z: Morrumbala Dist., near Vila Bocage, 3.x.1944, *Mendonça* 2338 (LISC). MS: Sussundega Dist., Lucite R., road between Gogoi and Dombe, 21.iv.1974, *Pope, Müller & Drummond* 1225 (K, LISC, SRGH).

Worldwide distribution except for the eastern and southern parts of South America and some islands (e.g. New Zealand). On still, often eutrophic waters, mostly lakes and large dams or amongst reeds; 100–1300 m.

Conservation notes: Widespread species; not threatened.

This is the most readily identifiable duckweed in the Flora area by virtue of the large, almost circular, frond with many roots. Landolt (1986) used abaxial surface colour of the frond as a key character to distinguish between *S. polyrrhiza* and *Landoltia punctata* (reddish vs. green). Examination of available material suggests that this is unreliable and best discarded as a field character given that the two species are readily separable on other characters.

Fig. 12.1.1. SPIRODELA POLYRRHIZA. 1a, ventral surface of sterile frond (× 8); 1b, cross-section of frond (× 8); 1c, resting bud (× 6); WOLFFIELLA HYALINA. 2a, fertile frond, dorsal view (× 12); 2b, frond in floating position (× 12); 2c, fruit and seed, median section (× 25); 2d, typical Lemnaceae pollen grains (× 160). WOLFFIELLA WELWITSCHII. 3a, fertile frond (× 12); 3b, stamen and pistil (× 20). WOLFFIA ARRHIZA. 4a, sterile frond with bud (× 20); 4b, fertile frond (× 20). 1 from *Brenan* 4995, 2a,b from *Chevalier* 1143, 2c,d after Daubs (1995), 3a from *Hall* 3036, 3b after Daubs (1965), 4 from *Hall* 3037. Drawn by Nigel Hepper. From F.T.E.A.

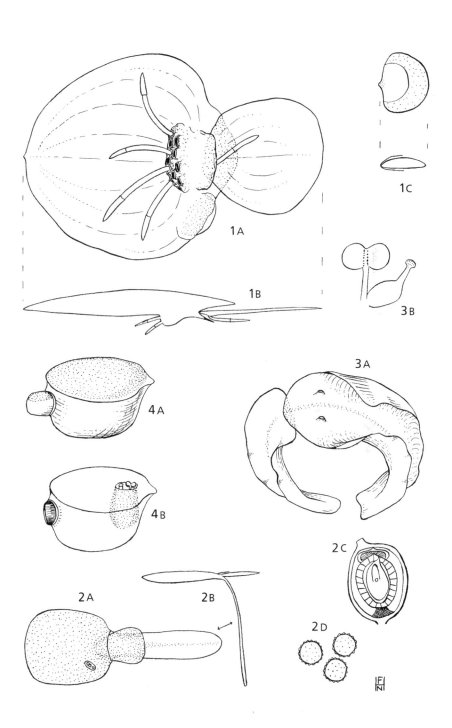

ARACEAE

2. LEMNA L.

Lemna L., Sp. Pl.: 970 (1753). —Daubs, Monogr. Lemnac.: 16–34 (1965). —Hartog & Plas in Blumea **18**: 360–364 (1970). —Landolt, Biosyst. Invest. Fam. Duckweeds, Family Lemnaceae **1**: 471–494 (1986). *Lenticula* Adans., Fam. Pl. **2**: 471 (1763).

Fronds ovate to lanceolate, floating on water surface, 1 to several cohering; frond with 1–5 'veins' and one type of crystal cell (raphides). Roots 1 per frond. Staminate flowers 2; anther bilocular with external theca situated above internal, dehiscing transversely. Pistil with 1 amphitropous or nearly orthotropous ovule. Fruit a dry pericarp with 1 ribbed seed.

Genus of 13 species distributed worldwide except in places too cold or dry. Two species in the Flora area.

The name *Lemna gibba* has often been used in checklists from the Flora area, but no specimens to confirm this are available. It is possible that this species does occur here as it is recorded throughout Tanzania. More collections are necessary to confirm this.

Root sheath not winged; root often exceeding 3 cm, root tip mostly rounded; frond usually tinged reddish or with reddish spots . . . . 1. *minor*
- Root sheath winged at base; root never exceeding 3 cm, root tip sharply pointed; frond lacking reddish coloration . . . . 2. *aequinoctialis*

1. **Lemna minor** L., Sp. Pl.: 970 (1753). —N.E. Brown in F.C. **7**: 41 (1897); in F.T.A. **8**: 202 (1902). —Hepper in F.W.T.A., ed.2 **3**: 129 (1968). —Hepper in F.T.E.A., Lemnaceae: 4 (1973). —Cook, Aq. Wetl. Pl. Sthn. Africa: 159 (2004). Type: Europe, LINN (BM lectotype), designated by Landolt (1986: 442). FIGURE 12.**1.2.1**.

Frond ovate, 1.5–2 times as long as wide, 1–8(10) × 0.6–5(7) mm, usually tinged reddish or with reddish spots. Root often exceeding 3 cm; root sheath not winged, root tip mostly rounded. Inflorescence produced occasionally. Pistil with 1 amphitropous ovule, style 0.1–0.15 mm long. Fruits rarely produced, 0.8–1 mm in diameter, with 1 seed. Seeds 0.7–1 × 0.4–0.6 mm, with 10–16 ribs, remaining within fruit at maturity.

Mozambique. GI: Homoine Dist., Inhanombe, 25.ii.1955, *Exell & Mendonça* 570 (BM, LISC, SRGH) [mixed collection with *Wolffiella denticulata* and *Wolffia arrhiza*].

Also found in North America, Europe, Africa and W Asia, introduced into S Australia and New Zealand; possibly native to southern Africa. In a wide range of aquatic habitats, often in somewhat eutrophic waters; c.50 m.

Conservation notes: Not threatened, but within the Flora area known only from one locality.

2. **Lemna aequinoctialis** Welw. in Ann. Cons. Ultram., parte não official, sér.1 [Apontamentos phytogeographicos] **55**: 578 (1859). —Brown in F.T.A. **8**: 203 (1902). —Wild in Kirkia **2**: 29 (1961).—Cook, Aq. Wetl. Pl. Sthn. Africa: 158 (2004). Type: Angola, Luanda, 10.i.1858, *Welwitsch* 206 (STU lectotype, BM, G, K), lectotypified by Landolt (1986). FIGURE 12.**1.2.2**.

Lemna angolensis Hegelm. in J. Bot. **3**: 112 (1865). Type: Angola, Luanda Dist., 10.i.1858, *Welwitsch* 206 (BM holotype, K).
Lemna paucicostata Engelm. in Gray, Man. Bot., ed.5: 681 (1867). Types from many countries.
Lemna perpusilla sensu Wild in Kirkia **2**: 29 (1961). —Hartog & Plas in Blumea **18**: 363 (1970). —Gibbs Russell in Kirkia **10**: 462 (1977). —Hepper in F.T.E.A., Lemnaceae: 4 (1973), non Torrey.

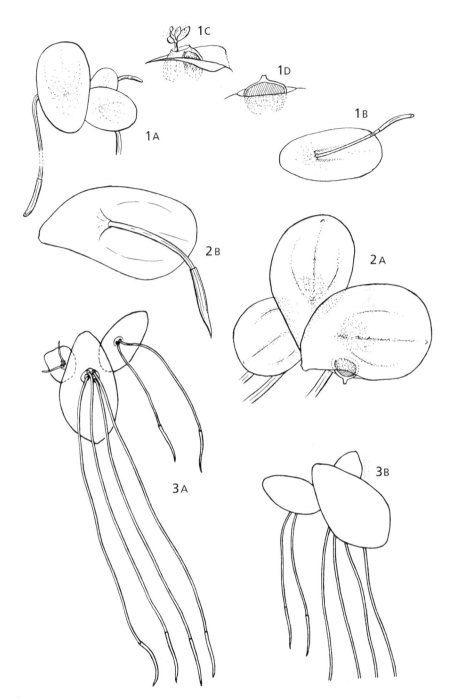

Fig. 12.1.**2**. LEMNA MINOR. 1a, sterile fronds (× 8); 1b, ventral surface of frond (× 8); 1c, dehisced anthers (× 27), 1d, fruit (× 27). LEMNA AEQUINOCTIALIS. 2a, fronds with fruit (× 8); 2b, ventral surface of frond (× 8). LANDOLTIA PUNCTATA. 3a, frond viewed from below (× 7); 3b, frond from above (× 7). 1a from *Thomas* 2410, 1b–d from *Melville* s.n., 2 from *Richards* 21000, 3 from *Ellis* 57610. Drawn by Nigel Hepper (1–2) and F. Crozier (3). From F.T.E.A. and Flore des Mascareignes.

ARACEAE

Frond ovate, 1–3 times as long as wide, 1–6.5 × 0.8–4.5 mm, green. Root never exceeding 3 cm; root sheath winged, root tip sharply pointed. Inflorescence often produced. Pistil with 1 orthotropous ovule, style 0.05–0.2 mm long. Fruit often produced, 0.5–0.8 × 0.4–0.7 mm, with 1 seed. Seed 0.45–0.8 × 0.3–0.7 mm, with 8–26 ribs, falling from fruit at maturity.

Botswana. N: Central Dist., c.3 km from mouth of Nata R., 28.iv.1956, *Drummond* 5278 (K, LISC, SRGH). SE: Kgatleng Dist., Mochudi, 22.iv.1967, *Mitchison* A42 (K). **Zambia.** N: Nchelenge Dist., Lake Mweru, 10.iv.1974, *Richards* 9139 (BR, K). E: Chipata Dist., Chinzombo Research Station, 17.ii.1988, *Phiri* 1870 (K). S: Choma Dist., Muckle Neuk, c.19 km N of Choma, 28.ii.1954, *Robinson* 589 (K). **Zimbabwe.** N: Makonde Dist., Chinhoyi (Sinoia) Caves, Sleeping Pool, 29.iv.1980, *van Heerden* 33 (K, LISC, LMU, SRGH). W: Hwange Nat. Park (Wankie Game Res.), v.1954, *Maar* 46806 (K, SRGH). S: Gwanda Dist., Doddiburn dam on Mtshibizini R., 10.v.1972, *Pope* 716 (K, SRGH). **Malawi.** N: Mazimba Dist., Lake Kazuni, 6.ii.1970, *Pawek* 3354 (K). S: Machinga Dist., Lake Chilwa, 20.viii.1978, *Howard-Williams* 232 (SRGH). **Mozambique.** T: Cahora Bassa Dist., Rio Mucangádzi, c.16 km along new road from Songo–Maroeira–Rio Mucangádzi, 22.v.1972, *Pereira & Correia* 2777 (LMU). MS: Beira, Macúti, 23.iii.1960, *Wild & Leach* 5197 (K, SRGH).

Distributed worldwide in tropical and subtropical areas; introduced further north into temperate regions with a rice culture. On eutrophic waters, roadside pools, ponds, dams and ditches; 0–1200 m.

Conservation notes: Widespread species; not threatened.

3. **LANDOLTIA** Les & D.J. Crawford

Landoltia Les & D.J. Crawford in Novon **9**: 532 (1999).

Fronds with (3)5–7 veins, with several layers of air spaces, pigment cells, raphides and druses, surrounded at base by a bifid scale (prophyll) covering the place of root attachment; roots (1)2–7(12) all perforate the scale. Utricular scale-like leaf surrounding the flowers open at one side. Pistil with one amphitropous ovule or 2–5 anatropous ovules. Seeds 0.8–1 mm long, with 10–15 longitudinal ribs.

A monotypic genus originally from Australia and Asia, now with a near worldwide distribution.

Landoltia punctata (G. Mey.) Les & D.J. Crawford in Novon **9**: 532 (1999). Type: Guyana, Essequibo R., *Meyer* s.n. (not located) (see note in Landolt 1986: 440); neotype from Chile, Tiera del Fuego Is., Orange Harbour, 1838, Wilkes Expedition (US neotype, DS, GH, KANU, MO), neotypified by Thompson (Rep. Miss. Bot. Gdn. **9**: 28, 1898). FIGURE 12.1.**2**.3.

Lemna punctata G. Mey., Prim. Fl. Esseq.: 262 (1818).
Spirodela punctata (G. Mey.) C.H. Thomps. in Ann. Rep. Miss. Bot. Gard. **9**: 28 (1898). — Gibbs Russell in Kirkia **10**: 462 (1977). —Landolt, Biosyst. Invest. Fam. Duckweeds, Family Lemnaceae **1**: 440 (1986). —Cook, Aq. Wetl. Pl. Sthn. Africa: 160 (2004).

Frond ovate to lanceolate, 1.5–2 times long as wide, 1–8 × 1–5 mm. Roots 2–7. Inflorescence occasionally produced. Pistil with 1–2 amphitropous ovules; style 0.15–0.2 mm long. Fruit 0.8–1 × 1–1.2 mm, containing 1 (rarely 2) seeds. Seed 0.8–1 × 0.5–0.6 mm, with 10–15 ribs.

Zimbabwe. N: Makoni Dist., Lomagundi, Chinhoyi caves, 29.iv.1980, *van Heerden* in *SRGH* 265938 (SRGH). W: Bulawayo, Hillside dams, 28.i.1972, *Biegel* 3820 (SRGH). C: Goromonzi Dist., Ruwa R. at bridge on Harare–Mutare road, 20.xi.1970, *Biegel* 3418 (BR, K, LISC, MO, PRE, SRGH).

Originally native to South America, southern Africa, SE Asia and Australia. Introduced to North America, Italy, Egypt, Israel, New Zealand and various islands in the Indian, Pacific and Atlantic Oceans. In the Flora area only known from Zimbabwe. Still open water; c.1400 m.

Conservation notes: Introduced species; not threatened.

4. WOLFFIELLA (Hegelm.) Hegelm.

Wolffiella (Hegelm.) Hegelm. in Bot. Jahrb. Syst. **21**: 303 (1895). —Daubs, Monogr. Lemnac.: 34–41 (1965). —Hartog & Plas in Blumea **18**: 364–365 (1970). — Landolt, Biosyst. Invest. Fam. Duckweeds, Family Lemnaceae **1**: 532–543 (1986).

   Pseudowolffia Hartog & Plas in Blumea **18**: 365 (1970).

   Wolffiopsis Hartog & Plas in Blumea **18**: 366 (1970).

Fronds flat, orbicular to strap-shaped, floating on or below water surface, 1 to many cohering, 'veins' and roots absent, pigment cells present (sect. *Wolffiella*) or absent (sects. *Stipitatae* and *Rotundae*). Staminate flower solitary; anther unilocular, dehiscing terminally; pistil with 1 orthotropous ovule. Fruit a dry pericarp containing 1 nearly smooth seed.

A genus of nine species arranged in three sections (Landolt 1986) distributed in temperate and tropical Africa and America, introduced to India. Five species representing all three sections found in the Flora area.

Hartog & Plas (1970) segregated *Wolffiopsis* based on differences in frond and pouch shape and number of inflorescences – frond and pouch asymmetric with a solitary inflorescence (*Wolffiella*) contrasted with frond and pouch symmetric with 2 inflorescences (*Wolffiopsis*) – and *Pseudowolffia* on frond orientation (floating vs. submerged) and the width of the pouch appendage (wide vs. narrow). Landolt (1986: 531) demonstrated convincingly that these genera cannot be upheld on these or any other characters.

1. Frond floating on water surface, orbicular to ovate, sometimes polygonal; pigment cells lacking. 2
   − Frond floating below water surface, linear to saddle-like, the tip lowermost; pigment cells present 4
2. Lower wall of pouch with an elongated, downward-bent, ribbon-like appendage 1–8 mm long; frond with 2 papules on upper surface .3
   − Lower wall of pouch without an appendage; frond lacking papules . . . **1.** *rotunda*
3. Frond with distinct air spaces spread through ½ –⁴/₅ of surface; pouch appendage 0.6–1.8 mm wide . . . . **2.** *hyalina*
   − Frond with indistinct air spaces restricted to area around base; pouch appendage 0.2–0.5 mm wide . . . .**3.** *repanda*
4. Frond with 2–4 teeth at tip; inflorescence solitary per frond, rarely flowering. . . . . **4.** *denticulata*
   − Frond lacking teeth at tip; inflorescences 2 per frond, often flowering . . . . **5.** *welwitschii*

1. **Wolffiella rotunda** Landolt in Veröff. Geobot. Inst. ETH Stiftung Rübel Zürich **70**: 26 (1980). Type: Zimbabwe, Hurungwe Dist., Zambezi Valley, W end Kariba Gorge, 25.xi.1953, *Wild* 4264 (SRGH holotype, B, K, LISC, LISU, MO, PRE, S, SRGH).

Frond ovate, 1–3 × 0.9–3 mm, margins entire; frond with indistinct air spaces restricted to area around base. Pouch lacking an appendage. Fruit unknown.

Zimbabwe. N: Hurungwe Dist., Zambezi Valley, W end of Kariba Gorge, 25.xi.1953, *Wild* 4264 (B, K, LISC, LISU, MO, PRE, S, SRGH).

Only known from the type. Pan in mopane woodland; c.500 m.

Conservation notes: Known only from a single locality, dammed in 1958; possibly extinct as locality was destroyed. Best treated as Data Deficient.

2. **Wolffiella hyalina** (Delile) C. Monod in Mém. Soc. Hist. Nat. Afri. N., Hors sér.2: 242 (1949). —Hepper in F.W.T.A., ed.2 **3**: 127 (1968). —Cook, Aq. Wetl. Pl. Sthn. Africa: 161 (2004). Type: Egypt, near Cairo, n.d., *Delile* s.n. (MPU holotype, P). FIGURE 12.1.1.2.

Lemna hyalina Delile, Descr. Egypte, Hist. Nat. (Fl. Aegypt. Illust.): 75 (1813).
Wolffia delilii Schleid. in Linnaea **13**: 390 (1839). Type as for *W. hyalina.*
Wolffia hyalina (Delile) Hegelm., Lemnac.: 128 (1868).
Pseudowolffia hyalina (Delile) Hartog & Plas in Blumea **18**: 366 (1970). —Hepper in F.T.E.A., Lemnaceae: 8 (1973).

Frond ovate to trapezoid, 1–3 × 0.8–2 mm, with or without a few single-celled teeth along margin, distinct air spaces distributed over ½–⁴/₅ of surface. Pouch appendage 0.5–5 × 0.6–1.8 mm, ¾–5 times as long and ¾–4 times as wide as frond. Seeds 0.35–0.4 × 0.25–0.3 mm.

Zambia. N: Nchelenge Dist., Lake Mweru-Wantipa, 10.iv.1957, *Richards* 9139 (BR, K). **Malawi**. S: Machinga Dist., Lake Chilwa, 22.ix.1970, *Müller* 1694 (K, SRGH).

Widespread throughout the drier tropical and subtropical regions of Africa; introduced into India. On pools and in slowly-moving water; 600–900 m.

Conservation notes: Local and possibly threatened in the Flora area, but not threatened globally.

3. **Wolffiella repanda** (Hegelm.) C. Monod in Mém. Soc. Hist. Nat. Afr. N., Hors sér.2: 242 (1949). —Cook, Aq. Wetl. Pl. Sthn. Africa: 161 (2004). Type: Angola, Luanda, Bemposta, 1.iii.1854, *Welwitsch* 205 (STU holotype, C).

Wolffia repanda Hegelm. in J. Bot. **3**: 113 (1865).
Pseudowolffia repanda (Hegelm.) Hartog & Plas in Blumea **18**: 366 (1970).

Frond ovate, 0.5–1.8 × 0.4–1.2 mm, with several single-celled teeth along margin; indistinct air spaces restricted to base. Pouch appendage 3–8 × 0.2–0.5 mm, ½–7 times as long and ¼–½ times as wide as frond. Seeds 0.3–0.35 × 0.22–0.27 mm.

Botswana. N: Okavango, Ngamaga Is., 14.v.1976, *P.A. Smith* 1735 (SRGH).

Also in Angola. Not gregarious, usually with other floating plants in pools and lakes; c.1000 m.

Conservation notes: Endemic to Angola and Botswana; Data Deficient or possibly Near Threatened.

4. **Wolffiella denticulata** (Hegelm.) Hegelm. in Bot. Jahrb. Syst. **21**: 305 (1895). — Cook, Aq. Wetl. Pl. Sthn. Africa: 161 (2004). Type: South Africa, Cape Province, 1841, *Krauss* s.n. (STU holotype).

Wolffia denticulata Hegelm., Lemnac.: 133 (1868).

Fronds dimorphic; vegetative frond submerged, ribbon-like, 2–7 × 0.3–0.8 mm; reproductive frond floating, ovate, 2–4 × 0.8–1 mm; both types with 2–4 short triangular teeth at tip. Fruit unknown.

Mozambique. GI: Homoine Dist., Inhanombe, 25.ii.1955, *Exell & Mendonça* 570 (BM, LISC, SRGH) [mixed collection with *Lemna minor* and *Wolffia arrhiza*].

Also in South Africa. Not gregarious, usually with other floating plants in pools and lakes; c.50 m.
Conservation notes: Known only from Mozambique and South Africa; Data Deficient or possibly Near Threatened.

5. **Wolffiella welwitschii** (Hegelm.) C. Monod in Mém. Soc. Hist. Nat. Afr. N., Hors sér.2: 242 (1949).—Hepper in F.W.T.A., ed.2 **3**: 127 (1968). —Cook, Aq. Wetl. Pl. Sthn. Africa: 161 (2004). Type: Angola, Ambriz Dist., Quizembo, xi.1853, *Welwitsch* 209 (STU holotype, BM, C, G, K, L, M). FIGURE 12.1.1.3.

Wolffia welwitschii Hegelm. in J. Bot. **3**: 114 (1865).
Wolffiopsis welwitschii (Hegelm.) Hartog & Plas in Blumea **18**: 366 (1970). —Hepper in F.T.E.A., Lemnaceae: 7 (1973).

Frond tongue-shaped, saddle-like basally, 3–7 × 2.5–5 mm, with 2–4 short triangular teeth at tip. Inflorescence 2 per frond, each arising from a separate terminal median pouch, often flowering. Fruits produced often, 0.5–0.6 × 0.35–0.4 mm.

Botswana. N: Moremi Wildlife Reserve, Moanachira R., 14.v.1973, *P.A. Smith* 585 (K, SRGH). **Zambia**. N: Nchelenge Dist., lagoon at Chiengi, n.d., *Whellan* 1393 (BR, SRGH). **Zimbabwe**. W: Hwange Dist., Ngamo pan, 15.viii.1950, *Garley* in *SRGH* 29022 (SRGH). **Malawi**. S: Machinga Dist., sand bar between Lakes Chilwa and Chiuta, 1.viii.1969, *Williams* 180 (SRGH). **Mozambique**. M: Inhaca Is., Tivanini swamp, c.37 km E of Maputo, 18.vii.1958, *Mogg* 28066 (PRE, SRGH).
Widespread throughout tropical America, the Caribbean and tropical Africa, including South Africa. Not gregarious, usually with other floating plants in pools and lakes; 0–1000 m.
Conservation notes: Widespread species; not threatened.

5. WOLFFIA Schleid.

Wolffia Schleid., Beitr. Bot. **1**: 233 (1844). —Daubs, Monogr. Lemnac.: 41–49 (1965). —Hartog & Plas in Blumea **18**: 366–367 (1970). —Landolt, Biosyst. Invest. Fam. Duckweeds, Family Lemnaceae **1**: 544–551 (1986).

Fronds spherical to ellipsoid, 0.5–1 times as deep as wide, greatest width slightly below water surface; 'veins' and roots absent; pigment cells absent. Staminate flower solitary, unilocular; pistil with 1 orthotropous ovule. Fruit a dry pericarp containing 1 nearly smooth seed.

A genus of 9 species arranged in four sections (Landolt 1986) distributed worldwide in temperate, warm temperate and tropical areas. Two species from one section (*Wolffia*) found in the Flora area.

Frond with 10–100 stomata; upper surface bright green, not transparent . . . . . .
. . . . **1.** *arrhiza*
 – Frond with 1–15 stomata; upper surface transparent . . . . **2.** *globosa*

1. **Wolffia arrhiza** (L.) Wimm., Fl. Seychelles, ed.3: 140 (1857). —Wild in Kirkia **2**: 29 (1961). —Hepper in F.W.T.A., ed.2 **3**: 129 (1968). —Hepper in F.T.E.A., Lemnaceae: 8 (1973). —Cook, Aq. Wetl. Pl. Sthn. Africa: 160 (2004). Type not known, probably from Europe (see Landolt 1986: 454). FIGURE 12.1.1.4.

Lemna arrhiza L., Mant. Pl. Alt. **2**: 294 (1771).
Wolffia michelii Schleid., Beitr. Bot.: 233 (1844), invalid name. —Brown in F.T.A. **8**: 205 (1901). Type as for *W. arrhiza*.

Frond spherical to ellipsoid, 0.5–1 × 0.4–1.2 mm, 1–1.3 times as long as wide, 1.25–1.5 times as deep as wide; upper surface bright green. Inflorescence occasionally produced. Plants rarely fruiting; seeds 0.4–0.5 × c.0.4 mm;

Botswana. N: Chobe, Kachikabwe (Kachikau), n.d., *Edwards* 4382 (PRE). **Mozambique.** GI: Homoine Dist., Inhanombe, 25.ii.1955, *Exell & Mendonça* 570 [mixed collection with *Wolffiella denticulata* and *Lemna minor*] (BM, LISC, SRGH).

Found in temperate, subtropical and tropical regions of Europe, Western Asia and Africa; introduced to Brazil. Often forming large populations on water surface of pools, sheltered lakes and in ditches; 10–1000 m.

Conservation notes: Widespread; not threatened.

2. **Wolffia globosa** (Roxb.) Hartog & Plas in Blumea **18**: 367 (1970). —Cook, Aq. Wetl. Pl. Sthn. Africa: 160 (2004). Type not known (see Landolt 1986: 454).

Lemna globosa Roxb., Fl. Ind., ed.1832 **3**: 564 (1832).

Wolffia cylindracea Hegelm., Lemnac.: 123 (1868). Type: Angola, Banzade, Libongo, ix.1858, *Welwitsch* 212 (BM holotype, STU).

Frond ellipsoid, 0.4–0.8 × 0.3–0.5 mm, 1.3–2 times as long as wide, 1–1.5 times as deep as wide, upper surface transparent pale green. Inflorescence occasionally produced. Plants rarely fruiting; seeds not seen.

Zimbabwe. W: Hwange Dist., Ngwashla road, 16.ii.1956, *Wild* 4766 (BR, K, LISC, SRGH). S: Chiredzi Dist., Manjinji Pan, 31.iii.1971, *Taylor* 176 (K, LISC, PRE).

Found in tropical E Asia and Africa, including South Africa; probably introduced to southern North America. Often forming large populations on water surface of pools and sheltered lakes; 300–1000 m.

Conservation notes: In the Flora area known only from Zimbabwe. Widespread species; not threatened.

6. **STYLOCHAETON** Lepr.[1]

Stylochaeton Lepr. in Ann. Sci. Nat., Bot., ser.2 **2**: 184 (1834). —Mayo, Bogner & Boyce, Gen. Araceae: 151 (1997).

Stylochiton Schott, Aroideae: 10 (1855), orthographic variant.

Seasonally dormant herbs (evergreen in West Africa), stem an underground rhizome, horizontal to erect, sometimes stoloniferous, roots thick, spindle-shaped, often very fleshy. Leaves solitary to several, cataphylls often conspicuously mottled and apically auriculate, sometimes persisting as a fibrous mass. Petiole with a long sheath, sheath usually narrowing upwards or apically obtuse, rarely auriculate. Leaf blade lanceolate to ± rounded or cordate-sagittate to hastate-sagittate, base truncate, auriculate or cordate; primary lateral veins pinnate or mostly arising basally, higher order venation reticulate. Inflorescences 1–4, produced with or before leaves; peduncle short, borne at or below ground level; spathe tube with united margin in lower part, often inflated at extreme base; sometimes constricted between a lower and upper inflated zone, rarely entire spathe narrowly cylindric or conic; spathe blade lanceolate-elliptic, ± gaping or unexpanded with a longitudinal slit, often greatly thickened. Spadix usually sessile, rarely stipitate, shorter than spathe; female zone densely flowered, contiguous with male zone, or separated by an axis naked or bearing sterile or bisexual flowers; male zone fertile to apex. Flowers unisexual, surrounded by a single cup-like perigone. Male flowers often with thick fleshy margins of perigone, stamens 2–7; filaments filiform, long, rarely thickened apically, connective slender, sometimes slightly thickened, anthers dehiscing by longitudinal slits, central pistillode usually present. Female flowers in single

---

[1] By Anna Haigh & Josef Bogner.

whorl or in spirals; perigone usually greatly thickened, glandular or mealy on upper surface; ovary 1–4-locular, ovules 1 to many per locule, anatropous, placenta basal, parietal or axile; stylar region thick, ± cylindric, exserted beyond perigone; stigma capitate to broadly discoid and massive. Fruit a berry, borne at or below ground level in a globose to cylindric infructescence, often rugose, 1 to few seeded. Seeds ovoid to ellipsoid; testa thin, black. 2n = 28 or 56.

A genus of c.19 species endemic to tropical and SE subtropical Africa, with its centre of diversity in Tanzania; 6 species in the Flora area.

The inflorescence often appears before the leaves and is very inconspicuous, with its lower part underground and usually greenish. As a result herbarium specimens are often incomplete. More detailed field studies are needed.

1. Female flowers 20 or more 2
  – Female flowers less than 20. 3
2. Petiole-base with conspicuous ridged, transverse, maroon-purple barring; leaf-blade deeply sagittate, basal lobes subequal to median lobe; old cataphylls and petiole-bases not persistent. . . . . **6.** *cuculliferus*
  – Petiole-base not transversely barred, or if barred then not ridged; leaf-blade broadly hastate-sagittate, rarely cordate-sagittate or oblong-lanceolate with a truncate to emarginate base, basal lobes shorter than median lobe; old cataphylls and petiole-bases persisting as a conspicuous net-fibrous mass . . . . . **1.** *natalensis*
3. Leaf-blade subcircular, very shortly cordate with overlapping basal lobes; leaves appressed to soil surface . . . . **5.** *euryphyllus*
  – Leaf-blade various but not subcircular; leaves erect 4
4. Spadix with a c.1.9 cm long naked area between male and female zones; whole spathe limb twisted when dry; rhizome horizontal, slender, with long internodes . . . . **2.** *tortispathus*
  – Spadix with male and female zones contiguous, if a sterile zone present, only c.1 mm long or with sterile flowers; spathe limb not twisted or only at extreme apex; rhizome usually vertical, thick 5
5. Spathe narrowly cylindric, limb reduced, forming ¼ or less of total spathe length; margin of staminate perigone narrow, unthickened; leaf-blade linear-lanceolate to hastate, median lobe to 6 cm broad, blade glabrous . . . . **4.** *borumensis*
  – Spathe-limb more than ¼ of total spathe length; margin of staminate perigone broad, massively thickened; leaf blade hastate- to cordate-sagittate, median lobe usually more than 6 cm broad, blade shortly pubescent, especially on veins . . . . **3.** *puberulus*

1. **Stylochaeton natalensis** Schott, Aroideae: 10, t.14 (1855). Type: South Africa, Natal, Durban (Port Natal), n.d., *Gueinzius* in *Herb. Sonder* (S holotype, BM, K).

Rhizome ± vertical, to 15 cm long or more, 0.5–1 cm wide, surrounded by a persistent net-fibrous mass of old cataphylls and leaf-sheath bases; roots numerous, stout, tuberous, to 35 cm long. Cataphylls 4–15 cm long, acute, sometimes with small apical ovate-lanceolate projections, glabrous. Petiole 4–43 cm long, green, sometimes with purplish-green transverse markings at base, glabrous or shortly pubescent basally; sheath ± half petiole length, 3.5–18 cm long, narrowing gradually to apex; blade very variable, oblong to triangular in outline, sagittate, sagittate-hastate, hastate, cordate-sagittate, broadly cordate to ± truncate, 5–31 × 1.8–26 cm, glabrous, apex acute to acuminate or apiculate, basal lobes, if present, separated by broad sinus. Inflorescence appearing with leaves; peduncle 2.5–13 cm long, glabrous to densely short-pubescent. Spathe 4–23 cm long, greenish to yellowish or purplish on outer surface with ribbed veins; tube erect, 2–9 cm long, upper part subcylindric, 0.5–1.4 cm in diameter, lower part slightly inflated, ovoid-ellipsoid, 1.3–2.5 cm long, 0.7–2 cm in diameter; limb lanceolate, narrowly acute-acuminate, somewhat thickened, yellow or purplish within, margins revolute or not, 1.8–13.5

cm long, erect or forward-curving, equal to or shorter than tube. Spadix shorter than spathe-tube, 1.4–8 cm long; male zone cylindric, 1.2–6 × 0.5–0.7 cm, ± contiguous with female zone; female zone cylindric to ellipsoid-cylindric, 0.3–3 × 0.9–1.1 cm. Male flowers with 2–5 stamens; perigones ± congested, rhomboid to irregularly lobed, margins sometimes massively thickened, conspicuously rugose, sometimes with several erect teeth; filaments slender, filiform; central pistillode subcylindric-conic, overtopping or shorter than perigone-margins. Female flowers 20–40, spirally arranged; perigone ± regular, 4–6-sided, massively thickened and smooth (at least when young) or mealy on upper surface; ovary subglobose, pale green, 2-locular to incompletely 2-locular; ovules 1–2 per locule; placentas axile, in centre of septum near its apex; style pale green, cylindric, 2–3 mm long; stigma capitate-discoid, to 1 mm broad. Infructescence cylindric to subglobose, up to 4 × 3 cm.

1. Leaf blade triangular, sagittate, cordate or hastate; primary lateral veins emerging at 30–50° 2
 – Leaf blade oblong, margin nearly parallel, base truncate to weakly cordate; primary lateral veins emerging at angle of 15–30(40)° . . . **ii) subsp. *obliquinervis***
2. Leaf blade large, 29–31 × 24–26 cm, cordate-sagittate . . . . . . **iii) subsp. *maximus***
 – Leaf blade smaller, usually hastate or narrowly sagittate . . . . .**i) subsp. *natalensis***

i) Subsp. **natalensis**. —Obermeyer in Fl. Pl. Afr. **42**: t.1648 (1972). —Schott, Gen. Aroid.: t.68 (1858). —Wood, Natal Plants **3**: t.207 (1900).

   Stylochaeton gazensis Rendle in J. Linn. Soc., Bot. **40**: 220 (1911). Type: Mozambique, Madanda Forests, 5.xii.1906, *Swynnerton 717* (BM holotype, K).

   Stylochaeton hennigii Engl., Pflanzenr. **73** IV, 23f: 32, fig.3K-O (1920). Type: Tanzania, Lindi Dist., Tendaguru, 1910, *Janensch & Hennig* 99 (B holotype, K photo).

   Rhizome to 15 cm or more, surrounded by a persistent net-fibrous mass of old cataphylls and leaf-sheath bases; Cataphylls 8–15 cm long, acute. Petiole 14–43 cm long glabrous or shortly pubescent basally. Leaf blade 5–22 × 1.8–22 very variable in shape, usually hastate or narrowly sagittate. Primary lateral veins emerging at 30–50°. Peduncle 4–13 cm, glabrous to shortly and densely pubescent. Spathe 6–23 cm long, greenish to yellowish or purplish on outer surface with ribbed veins; tube erect, 3.5–9 cm long, upper part subcylindric, lower part slightly inflated, ovoid-ellipsoid; limb lanceolate, narrowly acute-acuminate, somewhat thickened, yellow or purplish within. Infructescence cylindric to subglobose, up to 4 × 3 cm

   Zimbabwe. E: Mutare Dist., Banti Forest Reserve, Chitora Farm, fl. 4.xi.1967, *Mavi 577* (K). **Malawi**. S: Zomba Dist., "B & EA", fl. 27.x.1987, *Tawakali & Thera* 1225 (K). **Mozambique**. N: Macomia Dist., Quirimbas Nat. Park, 12°07'S, 40°15'E, fl. 6.xii.2003, *Luke et al.* 9882 (EA, K, LMA, MO, NHT). Z: Lugela Dist., Namagoa Estate, fl. 26.x.1948, *Faulkner* 315 (K). MS: Chibuli, fl. n.d., *Honey* 781 (K, PRE). GI: Panda Dist., Rio Inhassune, between Mangorro & Panda, fr. 7.iv.1959, *Barbosa & Lemos* 8522 (LISC). M: Moamba Dist., road to Chinhanguanine–Boane, fr. 21.xi.1979, *Schäfer* 7075 (K).

   Also found in Tanzania and South Africa. Open woodland, *Brachystegia* woodland and damp areas on arable land; 50–1800 m.

   Conservation notes: Widespread; not threatened.

ii) Subsp. **obliquinervis** (Peter) Bogner & Haigh, comb. nov. Type: Tanzania, Kigoma Dist., Uvinza, E of Malagarasi R., 2.ii.1926, *Peter* 36138 (B holotype). FIGURE 12.1.3.

   Stylochaeton obliquinervis Peter in Nachr. Ges. Wiss. Göttingen, Math.-Phys. Kl. **1929**(3): 201 (1929, publ.1930).

   Rhizome 2–3 cm long, surrounded by a persistent net-fibrous mass of old cataphylls and leaf-sheath bases. Cataphylls to 6 cm long, sometimes with small apical ovate-lanceolate projections. Petiole 4–8.5 cm long, sometimes shortly pubescent. Leaf blade coriaceous, oblong, margin nearly parallel, base truncate to weakly cordate; 2–3 pairs of basal veins, primary lateral veins

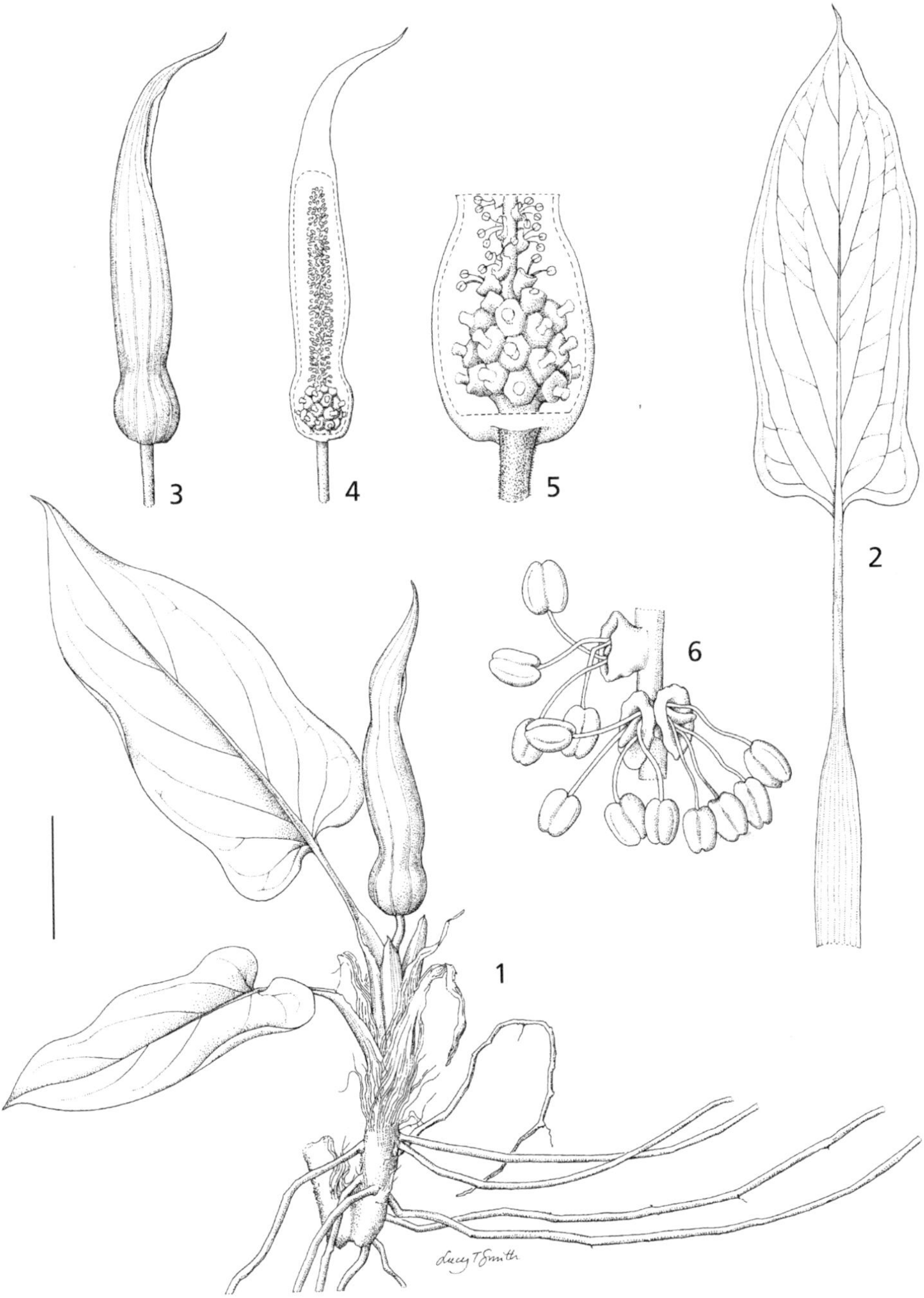

Fig. 12.1.**3**. STYLOCHAETON NATALENSIS subsp. OBLIQUINERVIS. 1, habit (× ³/₄); 2, leaf (× ³/₄); 3, inflorescence (× ³/₄); 4, inflorescence, spathe cut open (× ³/₄); 5, detail of spadix showing female flowers and lower male flowers (× 2¹/₂); 6, male flowers, attached (× 6). Scale bar: 1–4 = 2 cm, 5 = 7 mm, 6 = 2.5 mm. All from *Crawford et al.* 241. Drawn by Lucy Smith.

emerging at angle of 15–30(40)°. Peduncle 2.5–3.5 cm long, shortly and densely pubescent. Spathe 4–7.5 cm long, mid-green, weakly constricted above female zone of spadix; tube 2–4.5 × 0.5–1 cm, 0.6–1.2 cm at widest point; limb 1.8–3 cm long, shorter than tube, curved forwards, opening by a narrowly elliptic apical slit. Female flowers c.20.

Mozambique. N: Macomia Dist., NW of Namacubi Forest, W of Quiterajo, 11°44'02"S, 40°22'35"E, 90 m, fl. 26.xi.2008, *Crawford et al.* 241 (K, LMA).

Also in South Africa and Tanzania. Margins of open miombo woodland and seasonally waterlogged valleys; 90 m.

Conservation notes: Widespread distribution but not frequently collected; probably not threatened.

Mayo (1985) treats *S. obliquinervis* as a synonym of *S. borumensis*, but unfortunately the type of *S. obliquinervis* lacks an inflorescence. However, the leaves, stem and roots of *Crawford* 241 match the type well. *Crawford* 241 also has inflorescences with c.20 female flowers, a spathe with a larger opening and has persistent fibres formed from old cataphylls and leaf bases, suggesting a closer alliance with *S. natalensis*, hence the recombination made here. South African material (*Mauve & Ventner* 764, *Codd* 10125) supports this decision.

The leaves of this subspecies are very similar in shape and venation to *S. kornasii* Malaisse & Bamps, but this species only has a single whorl of female flowers, the stigma is enlarged and the filaments are strongly thickened apically.

iii) Subsp. **maximus** (Engl.) Bogner & Haigh, comb. nov. Type: Mozambique, Maputo Bay (Delagoa Bay), xi.1876, *Monteiro* 2 (P holotype, K). FIGURE 12.1.4.

　　Stylochaeton maximus Engl. in Bot. Jahrb. Syst. Pflanzen. **15**: 466 (1892). —N.E. Brown in F.T.A. **8**: 189 (1901). —Engler, Pflanzenr. **73** IV, 23f: 31 (1920).

Rhizome and roots unknown. Blade 29–31 × 24–26 cm, cordate-sagittate, basal half of each margin ± convex in outline, apex acute to broadly acuminate, apiculate at extreme tip; basal lobes separated by a hippocrepiform sinus; 3–4 pairs of basal veins, 2–3 pairs fused at base into a weak posterior rib, arcuate ascending to apex; 3–4 pairs of primary lateral veins arising at 45–50°, not easily differentiated from minor veins, also arcuate-ascending to apex. Spathe 13–20 cm long; tube 5.5–9 cm long, swollen at base, 2 cm at widest point; limb 7.5–12 × 1.5–3.5 cm wide, ± expanded, lanceolate, apex subulate. Female flowers 20–30; perigone ± regular, mealy on upper surface.

Mozambique. M: Maputo Bay (Delagoa Bay), xi.1876, *Monteiro* 2 (K, P).

Known only from the type and a sterile specimen from Zanzibar (*Holst* 2961). Growing in shade; c.50 m.

Conservation notes: Data Deficient.

Mayo (1985) treats *S. maximus* as a synonym of *S. natalensis* and comments that this is a distinct form. However the much larger and broader leaf blade suggest this would be better treated as a separate subspecies. As this taxon is known from only two specimens more field work is needed to support this.

2. **Stylochaeton tortispathus** Bogner & Haigh, sp. nov.[3] Type: Mozambique, Cabo Delgado, track to escarpment below Namacubi Forest, W of Quiterajo, 11°42'49.5"S, 40°23'46"E, 80 m, fl. 29.xi.2008, *Goyder et al.* 5072 (K holotype, LMA). FIGURE 12.1.**5**.

---

[3] **Stylochaeton tortispathus** sp. nova, a *S. bogneri* Mayo spatha tortiore constrictione unica instructa, spadicis axe inter zonas masculas femineasque longo, nudo differt. Type: Mozambique, Cabo Delgado, track to escarpment below Namaculi Forest, W of Quiterajo, 29.xi.2008, *Goyder et al.* 5072 (K holotype, LMA).

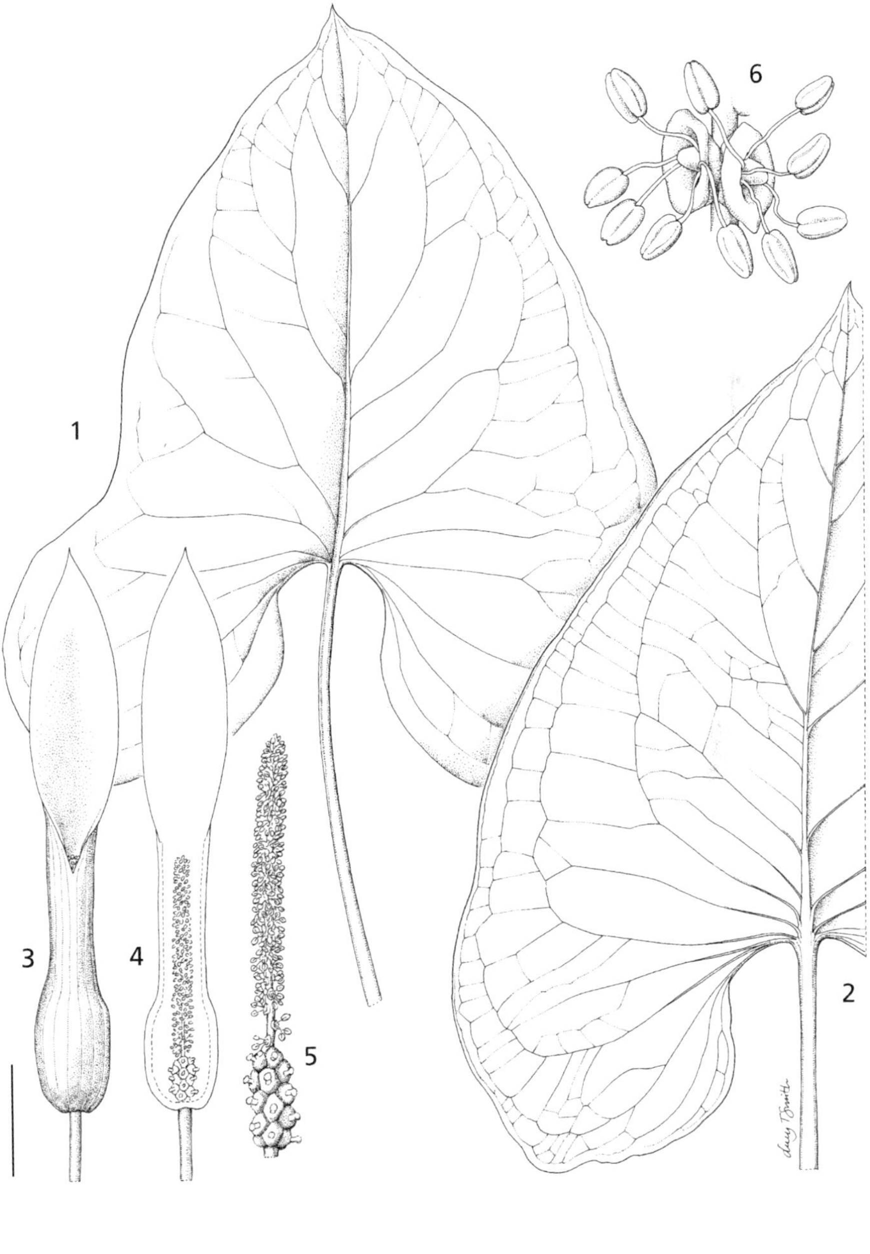

Fig. 12.1.4. STYLOCHAETON NATALENSIS subsp. MAXIMUS. 1, leaf, upper surface (× ⅓); 2, leaf, lower surface (× ⅓); 3, inflorescence (× ⅓); 4, inflorescence, spathe cut open (× ⅓); 5, spadix (× ¾); 6, male flowers, attached (× 6). Scale bar: 1–4 = 4 cm, 5 = 2 cm, 6 = 2.5 mm. All from *Monteiro* 2. Drawn by Lucy Smith.

Rhizome slender, 0.2 cm wide, internodes 2–3 cm long, roots slender, emerging from nodes. Cataphylls 4–6.5 cm long, acute, glabrous. Petiole 12–18 cm long, with purplish mottling at base, shortly pubescent basally, sheath to 3.5 cm long; blade broadly cordate, 12–16 × 10–17.5 cm, with a cuspidate apex to 0.7 cm long, anterior lobe convex, posterior lobes 6–9 cm long, rounded, spreading downwards to outwards, separated by a broadly parabolic to spathulate sinus; only 1 pair of primary lateral veins extending to apex, basal veins 3–5 pairs, collective veins formed from lower primary lateral veins and upper basal veins. Inflorescence solitary, appearing with leaves, borne at ground level in leaf litter. Peduncle shortly and densely pubescent, 3.8–5 cm long. Spathe 7–10 cm long, mottled reddish purple externally; tube 3–5.5 cm long, swollen at extreme base, followed by a strong constriction, swollen again above this; limb 3–4.5 cm long, strongly twisted on drying (unknown in life), apex subulate. Spadix 5.2 cm long; male zone 2.6 cm long, separated from female zone by a naked axis of 1.9 cm long; female zone globose, 0.7 cm long. Male flowers with 2–5 stamens; perigones congested, somewhat reduced, indistinct; stamens with filiform filaments 2–3 mm long, central pistillode not visible. Female flowers with c.8 flowers in 2 whorls; perigone ± regular, massively thickened and mealy on upper surface, pale; ovary subglobose, smooth on upper surface, unilocular with 8 free-central ovules, style stout; stigma pale globose. Fruit unknown.

Mozambique. N: Macomia Dist., W of Quiterajo, track to escarpment below Namacubi Forest, 11°42'49.5"S, 40°23'46"E, 80 m, fl. 29.xi.2008, *Goyder et al.* 5072 (K holotype, LMA).

Not known elsewhere. Locally common in humic soil in miombo woodland; 80 m.

Conservation notes: Known only from the type; Data Deficient.

This species appears to be closest to *S. bogneri* Mayo from Kenya and Tanzania owing to the slender horizontal rhizome, leaf shape, twisted spathe and number of female flowers. It differs from *S. bogneri* in the spathe having one strong constriction, not two, and having a long naked zone between the male and female flowers vs. a zone with sterile flowers. *S. tortispathus* also has a more strongly twisted spathe, hence the name.

3. **Stylochaeton puberulus** N.E. Br. in F.T.A. **8**: 188 (1901). —Engler in Pflanzenr. **73** IV, 23f: 31 (1920). —Peter in Nachr. Ges. Wiss. Göttingen, Math.-Phys. Kl. **1929**(3): 203 (1930). —Mayo in F.T.E.A., Araceae: 49 (1985). Type: Mozambique, Zambezi valley, Boroma (Boruma), near Tete, n.d., *Menyharth* s.n. (K holotype).

Rhizome ± vertical, 1–1.5 cm thick, often with distinct annular constrictions. Cataphylls 6–13 cm long, pubescent or glabrous, broadly oblong, apex rounded to auriculate-emarginate, with a central awl-like projection up to 1.2 cm long. Leaves solitary, sometimes 2; petiole 20–40 cm long, pubescent basally, sometimes with purplish or brown markings where it emerges from the soil; basal sheath 5–11 cm long, gradually narrowing upwards, sometimes obtuse to rounded; blade broadly triangular in outline, cordate-sagittate to cordate-hastate, usually pubescent beneath, often densely so on veins, 14–33 × 14–30 cm, sometimes broader than long, margins ± wavy, apex rounded and apiculate or shortly acuminate, basal lobes separated by a shallowly parabolic to oblong-spathulate sinus. Inflorescence appearing before leaves, semi-subterranean. Spathe erect, 3–5.5 cm long, lower half below ground, cream-coloured, pubescent to glabrous; tube 2–3.5 × 0.8–1.1 cm, base slightly inflated, 0.8–1.4 cm wide; limb 1–2.3 × 0.9–1.2 cm, substantially thickened, opening by a narrow lateral slit, margins not revolute, apex with a thick, conic, sometimes twisted cusp. Spadix much shorter than spathe, hidden within tube, 2.3–2.6 cm long; male zone subcylindric, 1.8 × 0.7 cm, contiguous with female zone or separated by a short naked interstice of c.1 mm; female zone subhemispherical, 0.8 × 0.9 cm. Male flowers with 3–4 stamens; perigones congested, rhomboid, margins rugulose, massively thickened; filaments filiform; central pistillode conic, overtopping perigone. Female flowers 12–18(30), spirally arranged; perigone subregular, minutely papillose; ovary flask-shaped, ± circular in section, 2-locular; locules each 3–4-ovulate with axile placentation; style very prominent, cylindric, 2.5–3 mm long; stigma subdiscoid, 1.5–2 mm broad. Berries white, shallowly papillose on upper surface, on a globose subterranean infructescence, c.3.5 cm wide.

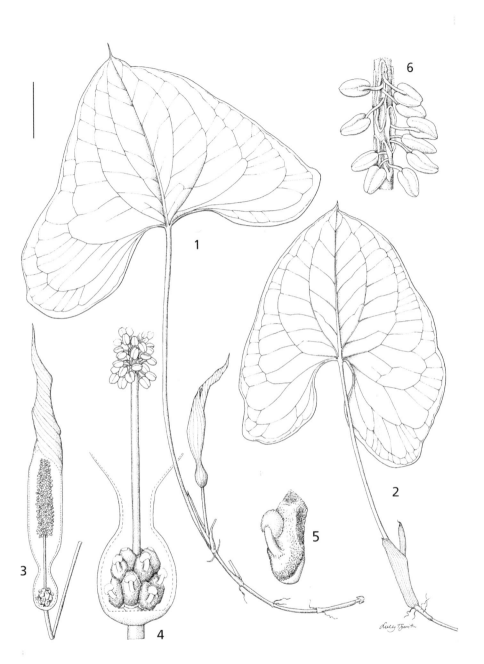

Fig. 12.1.5. STYLOCHAETON TORTISPATHUS. 1, habit with inflorescence (× ½); 2, habit with cataphylls (× ½); 3, inflorescence, spathe cut open (× ¾); 4, details of spadix, female flowers and lower section of male flowers (× 2); 5, female flower (× 7); 6, male flowers, attached (× 7). Scale bar: 1,2 = 3 cm, 3 = 2 cm, 4 = 7 mm, 5,6 = 2 mm. All from *Goyder et al.* 5072. Drawn by Lucy Smith.

Zambia. W: Mwinilunga Dist., just S of Matonchi Farm, fl. 11.x.1937, *Milne-Redhead* 2712 (K). N: Chiengi Dist., 14 km NE of Chiengi, fl. 13.x.1949, *Bullock* 1258 (K). E: Lundazi Dist., 10 km on Mfuwe–Mwanya road, 13°07'23"S, 31°54'27"E, 30.xi.2004, *Bingham* 12832 (K). C: Mumbwa, 25.iii.1964, *Van Rensburg* 2885 (K). S: Kazungula Dist., Machili, fr. 14.xi.1960, *Fanshawe* 5880 (K). **Zimbabwe**. N: Hurungwe Dist., Musukwi (Msukue) R., fr. 19.xii.1953, *Wild* 4216 (K, PRE). W: Hwange Dist. (Wankie), fr. 8.ii.1953, *Levy* s.n. (PRE). C: Chegutu Dist. (Hartley), iii.1929, *Eyles* 1355 (K). S: Mwenezi Dist., Nuanetsi Expt. Station, 15.i.1976, *Kelly* 552 (K, PRE). **Malawi**. N: Rumphi Dist, 1.6 km E of Rumphi, on E side of Chelinda R., 18.i.1976, *Pawek* 10728 (K). C: Salima Dist., 13 km S of Salima on main road, st. 8.iii.1982, *Brummitt, Polhill & Banda* 16397 (K). S: Chikwawa Dist., Lengwe Game Reserve, 13.xii.1970, *Hall-Martin* 1069 (K). **Mozambique**. N: Macomia Dist., Quiterajo, near end of airstrip, 11°44'52.7"S, 40°25'26.3"E, 24.xi.2008, *Goyder & Crawford* 5054a (K). Z: Mocuba Dist., Mocuba, fr. x.1948, *Faulkner* 324 (PRE).

Also in Kenya and Tanzania. Deciduous bushland, *Brachystegia* and mopane woodland and wooded grassland; 10–1400 m.

Conservation notes: Widespread; not threatened.

The semi-subterranean inflorescence makes this species difficult to collect in fertile condition; further complete collections are needed.

4. **Stylochaeton borumensis** N.E. Br. in F.T.A. **8**: 191 (1901). —Engler, Pflanzenr. **73** IV, 23f: 34 (1920). —Vollesen in Opera Bot. **59**: 108 (1980). —Mayo in F.T.E.A., Araceae: 51 (1985). Type: Mozambique, Zambezi Valley, Boroma, near Tete, i.1892, *Menyharth* 920 (K holotype).

Stylochaeton rogersii N.E. Br. in Bull. Misc. Inform., Kew **1912**: 283 (1912). Type: Mozambique, Nhamatanda (Vila Machado), fl. 18.ix.1911, *Rogers* 4500 (K holotype).

Rhizome vertical or horizontal, 0.5–1 cm thick. Cataphylls up to 12 cm long, white to pale green, usually densely mottled purple or greenish purple in irregular transverse bars, not persistent, apex acute. Leaves several; petiole 4–40 cm long, basal half mottled like cataphylls; basal sheath up to half total petiole-length, rarely more, gradually narrowing towards apex; blade very variable in shape, sometimes linear to ovate-elliptic with rounded to emarginate base, rarely broadly cordate, more often sagittate to hastate; 8–15 × 2–18 cm; posterior lobes variable, very short to as long as anterior lobe, 0.3–6 cm broad. Inflorescence appearing before leaves, usually present as leaves emerge, borne at or near soil-surface. Peduncle 1.5–6.5 cm long. Spathe glaucous green to pale yellow-green, often purplish brown towards apex and/or at base, or mottled purplish throughout, 4.5–15 × 0.5–1.5 cm, cylindric; apex erect, acuminate, with oblique ± circular opening usually less than quarter of total spathe length, margins not revolute or conspicuously thickened. Spadix hidden within spathe-tube, 2.5–9 cm long; male zone 0.7–0.9 times total length of spadix, cylindric, separated from 0.3–1 cm long female zone by 0.2–1 cm interstice bearing few sterile flowers. Male flowers with 3–4 stamens; perigones congested or distant, irregular or rhomboid with shallow unlobed to weakly 4-lobed margins or subrectangular with deeply 3–4-dentate margins; filaments somewhat to greatly thickened towards apex; central pistillode lacking. Female flowers usually 5–10, arranged in 1–2 whorls; perigone sac-like, oblique, surface mostly covered by minutely papillose, thick glandular area; ovary unilocular; ovules ± 20 on single basal placenta; style exserted, ± parallel to spadix-axis; stigma discoid to subglobose, to 3 mm wide. Berries (seen only in immature stage in Kenyan material) 8–10, grey-green, with persistent style-base and several oblong seeds. (Flower details taken from F.T.E.A.)

Zambia. N: Mpika Dist., N Luangwa Nat. Park (Luangwa Valley Game Reserve), c.600 m, fl. 17.xii.1966, *Prince* 34 (K, PRE). E: Mambwe Dist., Luangwa valley, c.15 km along Mfuwe–Chipata road, foot of escarpment, S13°17'16"S, 32°00'55", 580 m, fl. 6.xii.2004, *Bingham* 12839 (K). **Mozambique**. N: Macomia Dist., W of Quiterajo, below NW escarpment of Namacubi Forest, 11°42'53"S, 40°23'38"E, 5 m, fl. 29.xi.2008, *Goyder et al.* 5073 (K). Z: Lugela Dist., Namagoa and neighbourhood, fl. 23.x.1948,

Faulkner 329 (K). T: Moatize Dist., Boroma, Zambezi R., fl. i.1892, *Menyharth* 920 (K). MS: Nhamatanda Dist., Nhamatanda (Vila Machado), fl. 18.ix.1911, *Rogers* 4500 (K). Also in Kenya and Tanzania. Deciduous bushland and mopane woodland; 5–600 m. Conservation notes: Widespread; probably not threatened.

Stylochaeton obliquinervis was a synonym of this species in F.T.E.A. but is here placed as a subspecies of *S. natalensis*.

5. **Stylochaeton euryphyllus** Mildbr. in Notizbl. Bot. Gart. Berlin-Dahlem **13**: 410 (1936). —Mayo in F.T.E.A., Araceae: 54, fig. 13 (1985). Type: Tanzania, Lindi Dist., Lake Lutamba, 12.xii.1934, *Schlieben* 5719 (B holotype, BM, BR, G).

Rhizome vertical, 0.5–0.7 cm thick, with tuberous roots (in Tanzania). Cataphylls 4–7 cm long, oblong, subacute. Leaves 1–2; petiole subterranean, 3.5–7 cm long; basal sheath well developed, 3–6.5 cm long, nearly reaching petiole apex; blade subcircular, shortly and narrowly cordate, 6–21 × 6–19 cm, adpressed to soil-surface, upper surface bright green to dull bluish grey-green with yellowish to whitish venation, beneath paler and glossier with darker green venation, apex rounded, mucronate, basal lobes very short, ± overlapping, margin whitish-membranous. Inflorescence 1–2, appearing with leaves; base of spathe-tube ± subterranean. Peduncle short, subterranean. Spathe erect, 7–12 cm long, pale green to yellowish green; tube 2.5–6 × 0.5–0.8 cm, cylindric, hardly or not inflated at base; limb elliptic-lanceolate, 4.5–7 × 1.3–1.7 cm, ± spreading, somewhat thickened, greenish yellow on inner surface, acute-acuminate with twisted subulate cusp. Spadix subequal to or shorter than spathe-tube, 3–4 cm long; male zone cylindric, 2.5 × 0.5 cm, separated from female zone by a 1–4 mm long interstice with few sterile flowers or naked; female zone hemispherical, 0.4 × 0.6 cm. Male flowers cream-coloured with 3–5 stamens; perigones congested, rhomboid-polygonal, margins massively thickened, rugulose; filaments slender; central pistillode subconic, projecting above perigone margins. Female flowers 4–12, spirally-arranged in ± 2 whorls, or forming a single basal whorl; perigone regular, polyhedral, cream-coloured, massively thickened in upper part; ovary subglobose, smooth on upper surface, 2-locular, each locule with 1–2 axile ovules, rarely unilocular with up to 5 free-central ovules; style subcylindric, swelling apically; stigma discoid. Berries subglobose-polyhedral, 6 × 5–7 mm, upper surface papillose, 1–2-seeded; style-base persistent as blunt apical projection. (Flower and fruit details from F.T.E.A.)

Mozambique. N: Mueda Dist., Mueda Plateau, fl. 13.xii.2003, *Luke et al.* 10111 (EA, K, MO, LMA); Macomia Dist., Mirepa Forest, W of Quiterajo, NW escarpment of Namacubi Forest, 11°43'27"S, 40°23'21"E, fl. 29.xi.2008, *Goyder et al.* 5074 (K).

Also in Tanzania (T8). Forest floor and miombo woodland; 20–330 m. Conservation notes: Restricted distribution in Tanzania and only recently found in N Mozambique; probably Near Threatened.

6. **Stylochaeton cuculliferus** Peter, Nachr. Ges. Wiss. Göttingen, Math.-Phys. Kl. **1929**(3): 201 (1929, publ. 1930). —Mayo in F.T.E.A., Araceae: 51 (1985). — Lecron & Malaisse in Bull. Jard. Bot. Nat. Belg. **67**: 316, fig.9 (1999). Type: Tanzania, Tabora Dist., W of Usinge, Kombe, 26.i.1926, *Peter* 35590 (B holotype).

Rhizome vertical, 1–1.5 cm thick. Cataphylls 9–14 cm long, distinctly transversely barred with purplish barring and distinctly transversely ridged, narrowing apically, sometimes with terminal projection to 0.5 cm long (projection longer, stipitate and lanceolate in Tanzanian material). Petiole 13–26(40) cm long; basal part marked and ridged as cataphylls; basal sheath 2–9 cm long, rarely weakly auriculate or narrowing gradually upwards; blade narrowly triangular to ovate-elliptic in outline, deeply sagittate, 11–24 × 4–14 cm, acute, apiculate, basal lobes almost as long as anterior lobe, retrorse, acutely tipped, separated by narrowly triangular sinus. Inflorescence appearing before leaves, often paired, held at ground level. Peduncle mostly subterranean, slender, 4–11 cm long, much longer than spathe. Spathe pale green to brownish, erect, 4–5 cm long; tube 1–2.7 × 0.4–0.5 cm in upper narrower part; basal part inflated, obovoid,

1.5–1.8 × 1–1.2 cm; limb ovate-elliptic, 1.8–2.5 × 0.5–0.9 cm, massively thickened, margins not revolute, opening reduced to narrow lateral slit, apex with conic twisted cusp. Spadix 2.6–3.7 cm long, longer than spathe-tube, apex exposed within spathe limb; male zone cylindric, 1.5–1.9 × 0.4–0.5 cm, contiguous with female zone; female zone cylindric, 1.4 × 0.6 cm with 0.4 cm long naked basal stipe. Male flowers with 3 stamens; perigones congested, rhomboid, margins reddish-brown, substantially thickened; filaments short, slender; central pistillode white-translucent, hardly overtopping perigone-margins. Female flowers 27–30, spirally arranged; perigone ± regular, upper surface covered by glandular thickening, anticlinal walls reddish-brown speckled; ovary oblong and ± flattened on upper surface, unilocular; ovule solitary on parietal placenta near apex of side wall; style c.1 mm long; stigma subdiscoid, 1 mm broad. (Flower details from F.T.E.A.) Infructescence underground, subglobose, c.1.6 × 1.4 cm. Berries whitish, c.5 mm wide, with style and stigma remains (details from Malaisse & Bamps, Bull. Nat. Plant. Belg. **63**: 69–79, 1994).

Zambia. E: Mambwe Dist., Luangwa valley, Chindeni Hills, 13°14'15"S, 31°44'15"S, 640 m, fl. 3.xii.2004, *Bingham* 12838 (K). **Malawi**. N: Karonga Dist., c.8 km from Karonga, fl. 9.i.1959, *Richards* 10588 (K).

Also in Tanzania (T4) and Congo (Katanga). In *Brachystegia* woodland and grassland; 470–650 m.

Conservation notes: Moderately widespread species; not threatened,

7. ZAMIOCULCAS Schott

Zamioculcas Schott, Syn. Aroid.: 71 (1856). —Mayo, Bogner & Boyce, Gen. Araceae: 146–149 (1997).

Seasonally dormant or evergreen succulent herbs with a short, very thick rhizome. Leaves few to many, erect, subtended by several cataphylls; leaflets deciduous during dormancy leaving a persistent petiole. Petiole terete, base greatly thickened and succulent, pulvinate at apex, sheath very short, inconspicuous; blade pinnatisect, leaflets deciduous, oblong-elliptic, thickly coriaceous, capable of rooting at base once shed and forming new plants; primary lateral veins of each leaflet pinnate, running into marginal vein, higher order venation reticulate. Inflorescences 1–2, at ground level, appearing with the leaves; peduncle very short. Spathe persistent, slightly constricted between tube and limb, tube convolute, limb longer than tube, expanded and spreading to horizontally reflexed at anthesis. Spadix sessile, ± equalling spathe; female zone subcylindric, separated from male zone by short constricted zone bearing sterile flowers; male zone cylindric, ellipsoid to clavate, fertile to apex. Flowers unisexual, with 4 tepals in 2 whorls, decussate, thickened apically. Male flowers: tepals subprismatic; stamens 4, shorter than tepals; filaments free, oblong, thick, ± flattened; anthers introrse, connective slender; thecae ovate-ellipsoid, dehiscing by apical slit; pistillode clavate, equalling tepals. Sterile flowers each consisting of 4 tepals surrounding a clavate pistillode. Female flowers: staminodes lacking, gynoecium equalling tepals; ovary ovoid, 2-locular, 1 ovule per locule, hemianatropous, funicle short, placenta axile, near base of septum; style attenuate; stigma large, discoid-capitate. Infructescence ellipsoid. Berries depressed-globose with furrow at septum, 1–2-seeded, surrounded by persistent tepals, white. Seeds ellipsoid, testa smooth, brown.

A monotypic genus endemic to E and SE Africa.

Zamioculcas zamiifolia (Lodd.) Engl., Pflanzenr. **73** IV, 23b: 305, fig.85 (1905). —Mayo in F.T.E.A., Araceae: 15 (1985). Type: Illustration of cultivated plant from Loddiges, probably collected by Forbes in Mozambique, Bot. Cab. **15**: t.1408 (1829). FIGURE 12.1.**6**.

    Caladium zamiifolium Lodd., Bot. Cab. **15**: t.1408 (1829).

    Zamioculcas lanceolata Peter in Nachr. Ges. Wiss. Göttingen, Math.-Phys. Kl. **1929**(3): 209 (1930). Type: Mozambique, Beira, near Dondo, 5.x.1925, *Peter* 31194 (B holotype).

Fig. 12.1.**6**. ZAMIOCULCAS ZAMIIFOLIA. 1, habit (× ¹/₃); 2, tuber and petiole bases (× ²/₃); 3, leaflet (× ²/₃); 4, inflorescence (× ²/₃); 5, spadix, spathe removed (× 1); 6, male flower from above (× 4); 7, male flower, side view with one tepal removed (× 4); 8, female flower from above (× 4); 9, female flower in longitudinal section showing ovules and placentation (× 4). 1 from photo *Peter* s.n., 2, 3 from *Bogner* s.n., cult. Kew 106-76.00787, 4,5 from *Bogner* s.n., 6–9 from *Bally* s.n. Drawn by Eleanor Catherine. From F.T.E.A.

Rhizome up to 20 cm long and 10 cm in diameter, tough, woody. Leaves fleshy, glabrous. Petiole green with darker green to purplish transverse blotches, 23–35 cm long, 1–2 cm wide near base; blade 19–54 cm long; leaflets 5–9 per side, subopposite, distant, ovate to elliptic to obovate, sometimes narrowly-elliptic, dark glossy green, 5–12 × 1.3–4 cm, shortly acuminate, sessile or shortly petiolulate, articulated to rachis, cuneate to rounded basally; rachis terete, marked like petiole. Inflorescence subtended by cataphylls. Peduncle 6–10 × 0.5–1 cm, erect at first, recurving strongly in fruit, pushing infructescence into ground-litter. Spathe persistent, 5–9 cm long, coriaceous; tube shortly cylindric to ellipsoid, 1–2.2 × 1–2 cm, green on outer surface; limb broadly oblong-ovate, 5–6 × 3–5.5 cm, rounded and usually cuspidate apically, pale green to whitish, yellow or light brown. Spadix 5–7 cm long; female zone shortly cylindric-ellipsoid, 1–2 × 0.7–1.7 cm; male zone cylindric to clavate, dirty yellow, 4–5.5 × 1–1.5 cm, narrowed at base. Tepals white; stigmas yellowish. Berry white, to 1.2 cm broad, surrounded by persistent tepals, 1–2-seeded. Seeds brown, ellipsoid, c.0.8 × 0.5 cm broad.

Zimbabwe. E: Chimanimani Dist., Rusitu (Lusitu) R. near junction with Haroni R., st. 25.xi.1955, *Drummond* 5027 (K, LISC, PRE). **Malawi**. S: Mulanje Dist., Lichenya forest, 610 m, st. n.d., *Chapman* 447 (K). **Mozambique**. MS: Beira, bordering N side of Macuti beach, st. 10.ix.1962, *Noel* 2480 (K, LISC). GI: Inhambane, Lakeside, st. 3.xii.1935, *Moss* s.n. (LISC). M: Inhaca Is., st. 3.iv.1964, *Mogg* 30888 (K).

Also from Kenya, Tanzania and South Africa (KwaZulu-Natal). Forest floor in shade, often on river banks in sand. Common, sometimes forming pure cover; 0–650 m.

Conservation notes: Widespread; not threatened. The Sabonet Red Data List (Golding 2002) gives this as VU D2 for Zimbabwe.

As no mature fruiting specimens have been seen from the Flora area, most of the fruit characters have come from F.T.E.A. specimens.

In the East Usambara Mts in Tanzania juice from the leaves is used medicinally (*Msele* 14342), and it is also used for uncertain ailments in the Mulanje region, Malawi (*Chapman* 447).

8. GONATOPUS Engl.

Gonatopus Engl. in de Candolle & de Candolle, Monogr. Phan. **2**: 208 (1879). —Bogner in Aroideana **1**: 72 (1979). —Mayo, Bogner & Boyce, Gen. Araceae: 149 (1997).
Heterolobium Peter in Nachr. Ges. Wiss. Göttingen, Math.-Phys. Kl. **1929**(3): 211, 221 (1930).

Small to large seasonally dormant herbs; stem subterranean, a subglobose tuber or a cylindric, horizontal rhizome. Leaf solitary, glabrous, rarely pilose or scabrous, subtended by several cataphylls. Petiole pulvinate at base or centrally; blade broadly triangular to oblanceolate, usually trisect, rarely not (*G. petiolulatus*), primary divisions trifid to trisect or pinnatifid, or pinnatisect to quadri-pinnatifid, ultimate lobes linear to broad-elliptic, often decurrent; primary lateral veins of each lobe pinnate, forming an arching submarginal collective vein, higher order venation reticulate. Inflorescences 1–4, appearing before or with leaves, subtended by several cataphylls; peduncles erect, very short to long. Spathe tube convolute, subglobose, cylindric or suburceolate, constricted between tube and limb, limb oblong to elliptic, reflexed at anthesis, marcescent. Spadix subequal to spathe, female zone subcylindric, separated from male zone by a very short, constricted zone of sterile flowers; male zone longer than female, cylindric to clavate, fertile to apex. Flowers unisexual, with 4(6) free tepals, fleshy, truncate to ± hooded. Male flowers with united stamens forming a tube surrounding a central, cylindric to clavate pistillode, often exserted above tepals at anthesis; connective slender, thecae dehiscing by apical pore. Female flowers usually lacking staminodes, occasionally 1 staminode present; ovary 2-locular, ovules 1 per locule, anatropous, placenta basal-axile; style thick, stigma large, discoid-capitate. Berries ovoid-ellipsoid, 1–2-seeded, red or orange to yellow, or whitish. Seeds ovoid-ellipsoid, testa thin, smooth, embryo large.

A genus of 5 species endemic to tropical and southern subtropical Africa, with a centre of diversity in Tanzania.

A recently collected sterile specimen from N Mozambique may represent a new taxon (Cabo Delgado Prov., Nhica do Rovuma, c.40 km W of Palma, 8.xii.2008, *Goyder et al.* 5093 (K, LMA)). The leaflets are linear-lanceolate with the basal node of the leaf naked, and with sparse long dark hairs on the petiole. Collections of fertile material are needed.

1. Petiole with an aerial, elliptically swollen, central pulvinus; inflorescences usually several together (1–5); plant usually robust. . . . . **1.** *boivinii*
- Petiole with a cylindric basal pulvinus at underground point of attachment to tuber; inflorescence usually solitary. 2
2. Leaf blade broadly oblanceolate, lacking a distinct aerial petiole; apical pairs of pinnae largest, basal pairs reduced to oblanceolate or linear long-stipitate projections; inner surface of spathe-tube widely distant from female flowers . . . . **4.** *petiolulatus*
- Leaf blade broadly triangular with distinct aerial petiole; apical pairs of pinnae smaller than basal pairs; inner surface of spathe-tube tightly convolute around female flowers 3
3. Lowermost node of leaf naked, lacking subsidiary leaflets; ultimate leaflets of pinnae elliptic to oblong-elliptic, usually not decurrent; spathe-limb 5.5–9 cm long; male zone of spadix (0.8)1.2–1.9 cm in diameter; tuber usually subglobose or turnip-shaped . . . . **2.** *clavatus*
- Lowermost node of leaf usually with subsidiary leaflets; ultimate leaflets of pinnae linear, rarely broadly elliptic, usually decurrent; spathe-limb 2.5–5 cm long; male zone of spadix 0.8 cm wide; tuber usually cylindric and rhizomatous . . . . **3.** *angustus*

1. **Gonatopus boivinii** (Decne.) Engl. in de Candolle & de Candolle, Monogr. Phan. **2**: 209 (1879).—Mayo in F.T.E.A., Araceae: 10 (1985). Type: Zanzibar, n.d., *Boivin* s.n. (P holotype).

 Zamioculcas boivinii Decne. in Bull. Soc. Bot. France **17**: 321 (1870).

Robust herb. Tuber depressed-globose, 5–10 cm in diameter. Shoot subtended by several oblong, cream, purple-flecked cataphylls to 22 cm long. Leaves up to 150 cm tall; petiole terete, 25–75 cm long, erect, greyish to green, often blotched with purplish green; pulvinus conspicuous, central; blade trisect, primary divisions bipinnate, spreading, up to c.70 cm broad; terminal leaflets ovate to elliptic, sometimes lanceolate, 2–14 × 0.8–7 cm, green, paler beneath, acuminate, base acute to obtuse, usually shortly petiolulate sometimes sessile, not decurrent (terminal leaflets occasionally decurrent or partially united basally). Inflorescences 1–5, appearing just before leaf emergence. Peduncle 30–75 × c.1 cm, erect, greenish to cream, often with grey or purplish mottling. Spathe 10–21 cm long; tube ovoid or subglobose, tightly convolute around female flowers, constricted slightly at apex, 1.5–3 × 1.4–2.3 cm, darker than spathe limb, lined and speckled maroon to purple; limb oblong, 9.5–19 × 2–4.5 cm, long-cuspidate, widely spreading and reflexed, outer surface paler than tube, with similar markings, inner surface cream, sometimes speckled pink-brown. Spadix 8–15 cm long; female zone cylindric, 1.5–2.8 × 0.8–1.6 cm; male zone cylindric, 6.5–12.5 × 0.5–1.4 cm; intermediate sterile zone constricted, short, c.3 mm long. Male flowers with 4 tepals, cream to yellowish cream, filaments united, forming a tube surrounding central pistillode. Female flowers with 4 tepals, c.2.5 mm long; ovary 2-locular; placentation axile; stigma discoid, green, c.1.5 mm broad. Berry somewhat compressed laterally, obovoid with pronounced septal suture, c.1.3 × 1.4 cm, 2-seeded. Seeds obovoid-cylindric, subterete, smooth, 1 × 0.6 cm; raphe 0.8 cm long.

Zambia. N: Nchelenge Dist., Lake Mweru, fl. 13.ii.1957, *Fanshawe* 3955 (K).

E: Chipata Dist., Mwangazi valley, fl. 26.xi.1958, *Robson* 726 (K). **Zimbabwe**. E: Chimanimani Dist., above Haroni R. near confluence with Rusitu, 12.i.1969, *Biegel* 2814 (K). **Malawi**. S: Zomba Dist., Mpilise, Balaka, fl. 5.xii.1956, *Jackson* 2090 (K, LISC). **Mozambique**. N: Mandimba Dist., Mandimba, fl. 6.xi.1941, *Hornby* 3462 (K). Z: Lugela Dist., Namagoa, fl. x–xi.1943, *Faulkner* 195 (K, PRE). T: Cahora Bassa Dist., Songo, fl. 27.xi.1973, *Macêdo* 538 (LISC). MS: Chimoio Dist., roadside W of Bandula, 700 m, fl. 19.ix.1961, *Chase* 7559 (K, SRGH). M: Matutuine Dist., Bela Vista, Santaca, track to Frazão farm, fl. 13.xii.1967, *Gomes e Sousa* 5024 (PRE).

Also in Kenya, Tanzania, Congo and South Africa (KwaZulu-Natal). In moist forest, *Brachystegia* woodland and wooded grassland; 40–1000 m.

Conservation notes: Common and widespread; not threatened.

As no fruiting specimens were seen from the Flora area, fruit characters are taken from F.T.E.A. specimens.

2. **Gonatopus clavatus** Mayo in F.T.E.A., Araceae: 11 (1985). Type: Tanzania, Masasi Dist., Mbangala R., 15.xii.1955, *Milne-Redhead & Taylor* 7670 (K holotype). FIGURE 12.1.7.

Robust herb. Tuber subglobose to turnip-shaped, c.5 × 6 cm, or cylindric and up to 18 × 7.5 cm (in Tanzania). Leaves up to 1.2 m tall, glabrous; petiole 20–52 cm long, green with purplish tinges and greenish-white flecks; pulvinus basal, hidden by cataphylls; blade trisect, primary divisions bipinnate, up to c.60 cm broad, basal leaf node lacking subsidiary leaflets, terminal leaflets of pinnae (3)4–13 × (1)1.5–4.5 cm, elliptic to oblanceolate, acuminate, cuneate at base, mostly sessile, some petiolulate, terminal leaflets sometimes ± decurrent. Veins usually prominent on lower surface when dry. Inflorescence solitary, appearing before leaves. Peduncle 3–25 × 0.8–1.2 cm, glabrous to pubescent. Spathe 7.5–12 cm long; tube subglobose, constricted apically, 1.5–2 × 1.5–2.2 cm, reddish to deep wine red on inner surface; limb broadly elliptic to oblong-elliptic, 5.5–9.5 × 2.5–7.5 cm, long-cuspidate, reflexed at maturity, green with purplish tinge or reddish brown. Spadix 4.5–12 cm long; female zone ± cylindric, 1–2.5 × 0.8–1.7 cm; male zone clavate, sometimes subcylindric, 4.8–9.5 × (0.8)1.2–2 cm, creamy yellow, apex rounded, constricted at base, ± contiguous with female zone. Male flowers with 4 tepals, anthers slightly exserted from tepals at anthesis. Female flowers with 4 tepals, ovary ± laterally compressed, 2-locular; placentas at base of septum; stigma discoid, prominent, 2 mm in diameter. Fruit not seen.

Malawi. S: Mangochi Dist., E of Lake Malawi, 21 km N of Malindi, 19.xi.1977, *Brummitt et al.* 15108 (K). **Mozambique**. N: Monapo Dist., forest reserve of Sr. Wolf, 15.ii.1984, *Groenendijk et al.* 1059 (K). Z: Lugela Dist., Namagoa, fl. xi.1943, *Faulkner* 191 (K, PRE).

Also in Tanzania. In open *Brachystegia* woodland, disturbed woodland and stream margins; 50–500 m.

Conservation notes: Common and moderately widespread; not threatened.

Faulkner 329 and 496 (K) from Mozambique have scabrous peduncles and petioles, and smaller more membranous leaflets, but in other respects seem to match *G. clavatus*. More field studies are required to determine whether these should be recognised as a separate taxon.

Fig. 12.1.7. GONATOPUS CLAVATUS. 1, plant with leaf (× ¹/₆); 2, leaflets (× ²/₃); 3, plant with inflorescence (× ¹/₆); 4, inflorescence (× 1); 5, female zone of spadix, part of spathe-tube removed (× 1); 6, male flower from above (× 6); 7, tepal of male flower (× 8); 8, androecium (× 8); 9, pistillode of staminate flower (× 8); 10, pistillate flower from above (× 6); 11, tepal of female flower (× 8); 12, ovary, side view (× 8); 13, ovary in longitudinal section showing ovules and placentation (× 8); 14, ovary in transverse section (× 8). All from *Milne-Redhead & Taylor* 7670. Drawn by Ann Webster. From F.T.E.A.

3. **Gonatopus angustus** N.E. Br. in Bull. Herb. Boiss., sér.2 **1**: 778 (1901). —Mayo in F.T.E.A., Araceae: 12 (1985). Type: Mozambique, Tete, Boroma, iii.1891, *Menyharth* 922 (K holotype, WU).

Gonatopus rhizomatosus Bogner & Oberm. in Bothalia **12**: 251 (1977). Type: South Africa, Transvaal, 11.i.1961, *Codd* 7814 (PRE holotype, K, M).

Medium sized herb. Tuber horizontal and cylindric (especially in S Mozambique and South Africa), rarely subglobose to turnip-shaped, 4–17 × 1.5–4 cm. Leaves 20–65 cm tall, glabrous; petiole 10–37 cm long; pulvinus basal, cylindric, hidden by cataphylls; blade trisect, primary divisions pinnate to bipinnate, 15–30 cm wide; basal node of leaf usually with subsidiary leaflets or pinnae, ultimate leaflets 1–9 × 0.1–2(4.5) cm, linear to broadly elliptic, acuminate, ± narrowed basally, usually sessile or decurrent. Inflorescence usually solitary, appearing just before leaf. Peduncle 3–10 × 0.5–0.6 cm, elongating in fruit, terete. Spathe 3.5–6 cm long; tube broadly ovoid, constricted apically, 1.1–1.4 × 1.2–1.3 cm, outside brownish green, paler within; limb broadly elliptic, 2.5–5 × 2.3–3 cm, cuspidate, reflexed at maturity, pale brown with darker veins on outside. Spadix sessile, 3.9–5.3 cm long; male zone cylindric to somewhat clavate, 2.5–4 × 0.8 cm, white, apex rounded, constricted at base; sterile intermediate zone conic, 0.7–0.8 × 0.4–0.5 cm at narrowest point, white; female zone shortly subcylindric, 0.7–0.8 × 1 cm. Male flowers with 4(6) tepals, irregularly shaped, stamens sometimes only shortly united. Female flowers with 4 light green tepals; staminodes 0–1, subcylindric; ovary ± laterally compressed, 2-locular; placenta basal-axile; stigma prominent, discoid, 1–1.5 mm in diameter, dark green. Infructescence shortly cylindric; berry ovoid, 0.9–1.8 × 0.9–1.5 cm, 1–2-seeded, orange to yellow. Seed ovoid-elliptic, 0.5–1.2 × 0.3–0.7 cm.

Zimbabwe. E: Nyanga Dist., Honde R., fl. 15.xi.1960, *Wild* 5250 (LISC). S: Beitbridge Dist., 16 km S of Runde (Lundi) R., ii.1967, *Wild* 7610 (K, LISC, PRE). **Malawi**. S: Chikwawa Dist., Kapuchira (Livingstone) Falls on lower Shire R., fl. 15.vi.2007 (voucher from living collections), *Brummitt* 10016 (K). **Mozambique**. Z: Lugela Dist., Lugela, Moebede road, 23.iii.1949, *Faulkner* 398 (K). T: Cahora Bassa Dist., Songo, transmission station (Posto de Repetição) close to dam, fr. 5.ii.1973, *Torre, Carvalho & Ladeira* 19016 (BR, LD, LISC, WAG). M: Marracuene Dist., Marracuene, c.2 km towards Costa do Sol, fl. 13.xi.1980, *Nuvunga & Conjo* 385 (K).

Also in Tanzania and South Africa (former Transvaal, KwaZulu-Natal). In forest, often on sandy soil on river banks; 0–800 m.

Conservation notes: Widespread lowland species; not threatened.

The species has variable leaf morphology; the broader leaved form appears to be associated with shady conditions, but more field studies are required.

4. **Gonatopus petiolulatus** (Peter) Bogner in Aroideana **1**: 72 (1978). —Mayo in F.T.E.A., Araceae: 14 (1985). Type: Tanzania, Uzaramo Dist., near Msua, *Peter* 31907 (B† holotype). Neotype selected here: Tanzania, Uzaramo Dist., 14 km WSW of Dar es Salaam, Gongolamboto cemetery copse, 21.xi.1977, *Wingfield* 4439 (DSM, K photo).

Heterolobium petiolulatum Peter in Nachr. Ges. Wiss. Göttingen, Math.-Phys. Kl. **1929**(3): 211 (1929).

Heterolobium dilaceratum Peter in Nachr. Ges. Wiss. Göttingen, Math.-Phys. Kl. **1929**(3): 211 (1929). Types: Tanzania, Lushoto Dist., between Bamba and Mhinduro, *Peter* 24998 (B† syntype); Lushoto Dist., 27.ix.1918, *Peter* 25005 (B syntype).

Gonatopus latilobus K. Krause in Notizbl. Bot. Gart. Berlin-Dahlem **15**: 397 (1941). Type: Tanzania, Lindi Dist., c.20 km S of Lindi, near Mlinguru, 3.i.1935, *Schlieben* 5923 (B holotype).

Small to medium sized herb. Tuber subglobose, up to 8 cm in diameter (from F.T.E.A.). Leaves 22–40 cm long, shortly pilose to glabrous; petiole not distinctly developed; leaflets extending to ground-level, subterranean axis slightly thicker, forming a basal cylindric pulvinus; blade oblanceolate in outline, 8–12 cm broad at widest point, pinnate to bipinnate, pinnae in opposite pairs, basal pairs reduced to tufts of oblanceolate to linear, long-stipitate foliar projections;

leaflets glossy dark green on upper surface, minutely pilose below, shortly acuminate, decurrent or shortly petiolulate, usually with subsidiary foliar projections from base, terminal leaflets elliptic to oblanceolate, 4–8 × 1.8–3 cm. Inflorescence solitary. Peduncle 20–23 cm, erect, terete, minutely pilose. Spathe 5–10 cm long, sometimes minutely pilose externally; tube broadly suburceolate, only slightly constricted at mouth, distant (3–5 mm) from female flowers, 1.5–2.5 × 1.3–1.8 cm; limb elliptic-lanceolate, 3.5–7.5 × 2–2.6 cm, reflexed at anthesis, bright green, paler within, cuspidate, outer surface sometimes shortly pilose. Spadix 3.7–7.5 cm long; female zone cylindric, 0.8–1.6 × 0.5–0.8 cm; sterile intermediate zone 0.2–0.5 × 0.3–0.5 cm at narrowest part; male zone cylindric to clavate, 2.5–4.5 × 0.3–0.7 cm, cream, apex rounded, narrowing towards base. Male flowers 5 mm wide, with 4 tepals, filaments shortly united. Female flowers with 4 pale green tepals; ovary 2-locular; placentas at base of septum; stigma prominent, capitate-discoid, 1.5 mm in diameter. Fruit not seen.

Mozambique. N: Nangade Dist., Hunter's concession, between Pundanhar and Nangade, 40 m, fl. 17.xi.2009, *Goyder et al.* 6080 (K, LMA, LMU, P). Z: Lugela Dist., Namagoa and Moebede road, fl. 22.xi.1948, *Faulkner* 323 (K).

Also in Tanzania. In evergreen forest; 50–100 m.

Conservation notes: Only two specimens seen from the Flora area, possibly Vulnerable; widespread in Tanzania.

A small plant compared to most of the Tanzania specimens. More collections are needed to give a more complete picture.

9. ANCHOMANES Schott

Anchomanes Schott in Oesterr. Bot. Wochenbl. **3**: 314 (1853). —Mayo, Bogner & Boyce, Gen. Araceae: 218 (1997).

Large seasonally dormant herbs, often very robust; tuber small to gigantic, creeping obliquely or growing vertically. Leaves solitary, often gigantic, subtended by several basal cataphylls. Petiole erect, very long, terete, usually with short prickles, sheath very short. Leaf blade sagittate when juvenile, becoming trisect at maturity, primary divisions subdivided ± dichotomously or pinnately, secondary divisions irregularly pinnatifid, ultimate lobes very variable in size and shape, distal ones larger, trapezoid, apically broader, truncate or shallowly bifid, decurrent to sessile, proximal lobes ovate and acuminate; irregular submarginal collective vein sometimes present, higher order venation reticulate. Inflorescence solitary, usually appearing before the leaf. Peduncle usually with short prickles, shorter than petiole. Spathe erect, longer than or subequal to spadix, broadly ovate to narrowly oblong-lanceolate, boat-shaped, not constricted, convolute basally or not at all, apex sometimes arched over, marcescent. Spadix cylindric, female zone subequal to male zone or much shorter, male zone contiguous with female, fertile to apex. Flowers unisexual, perigone absent. Male flowers: anthers sessile, connective slender below, thickened and dilated apically, thecae ovate-oblong, opposite, dehiscing by apical slit. Female flowers: ovary 1-locular, ovule 1, erect, anatropous, funicle very short, placenta basal, style shortly conic or absent, sometimes strongly deflexed towards spadix base, stigma either 2-lobed and reniform to V-shaped or discoid or depressed-globose. Berries large, oblong-ellipsoid, fleshy, in a cylindric spike, red, purplish or partly white. Seeds obovoid to oblong-ovoid, testa very thin, smooth, transparent, embryo large, green, endosperm absent.

A genus endemic to tropical Africa with 7–8 species, only 2 in the Flora area.

Further field studies are required to improve identification of available sterile material, which currently lacks useful diagnostic characters.

Spathe broadly ovate to subcircular; female zone of spadix equalling or longer than male zone; in N Mozambique . . . . **1.** *abbreviatus*
– Spathe oblong-ovate; female zone of spadix shorter than male zone; in N & W Zambia . . . . **2.** *difformis*

1. **Anchomanes abbreviatus** Engl. in Bot. Jahrb. Syst. **36**: 237 (1905). —Engler, Pflanzenr. **73** IV, 23c: 52 (1911). —Mayo & Bogner in F.T.E.A., Araceae: 25 (1985). Type: Tanzania, Lushoto/Tanga Dist., Umba, *Kassner* 92 (BM† holotype). FIGURE 12.1.**8**.

Perennial herb. Tuber ± globose to subcylindric, growing vertically, 1–3 cm in diameter or more, roots thick, fleshy, 3–4 mm wide. Petiole 100–200 × 1.2 cm or more, marked with elongate creamish spots, with 1–5 mm long greenish cream downward pointing prickles; leaf blade 120–180 cm wide, dark glossy green. Peduncle 30–90 cm long, similar in colour to petiole, armed with prickles. Spathe broad-ovate, 9 × 2 cm, apex rounded to acute with cuspidate tip. Spadix cylindric, 4.8 × 1 cm, much shorter than spathe; male zone 2.4 cm long, apex rounded; female zone 2.4 cm long, ± half total spadix length. Male flowers 1 mm broad. Female flowers 5 × 3 mm, broadly subcylindric; stigma sessile, circular, whitish, hardly prominent, 2 mm wide. Berry ellipsoid, 2.7 × 1.6 cm. Seed subellipsoid, 2.3 × 0.9–1.3 cm.

Mozambique. N: Macomia Dist., 24.5 km N of Macomia on road to Diaca, fl. 2.xii.2008, *Burrows & Burrows* 10917 (K).

Also in Kenya and Tanzania. Dense deciduous woodland to open *Brachystegia* woodland; c.350 m.

Conservation notes: In the Flora area known from only one specimen in coastal woodland in N Mozambique; probably Near Threatened.

Anchomanes abbreviatus in the Flora area is geographically disjunct from *A. difformis*.

2. **Anchomanes difformis** (Blume) Engl. in de Candolle & de Candolle, Monogr. Phan. **2**: 304 (1879). —Hepper in F.W.T.A., ed.2 **3**: 121 (1968). —Mayo & Bogner in F.T.E.A., Araceae: 25 (1985). Type: Nigeria, Owerri, 1789, *Palisot de Beauvois* (G holotype).

Amorphophallus difformis Blume in Rumphia **1**: 149 (1937).

Anchomanes welwitschii Rendle in Hiern, Cat. Afr. Pl. Welw. **2**: 88 (1899). Types: Angola, Pungo Andongo, x.1856–iii.1857, *Welwitsch* 226 (BM syntype); Ambaca, x.1856, *Welwitsch* 225 (BM syntype).

Perennial herb. Tuber massively thickened, 6–20 cm in diameter, growing horizontally, with distinctive annular leaf-base scars, growing point situated obliquely, partly emerging from ground. Cataphylls oblong, subtending petiole. Petiole up to 300 × 1–4 cm, dark to purplish green with whitish speckling, with 2–3 mm long, outward-pointing, whitish-green prickles; blade up to 150 cm wide, dark green. Peduncle 50–180 × 0.6–1 cm, markings similar to petiole, smooth to prickly. Spathe oblong-ovate, 14–30 × 3–5 cm, acuminate. Spadix subcylindric, 12–30 cm long, shorter than spathe; male zone 5.5–12.5 × 0.7–1 cm, tapering towards blunt apex; female zone 1.5–3.5 × 0.8–1.5 cm. Stamens prismatic, 1.5–2.5 × 1.5 mm, creamy-white. Female flowers c.4 mm long; ovary 1.5–2.5 mm wide; style up to 2 mm long, sometimes ± sessile, narrowing apically, downward pointing; stigma reniform to discoid. Berries oblong-ellipsoid, red to purplish or white, 0.9–2 × 0.5–1 cm, rounded, 1-seeded. Seed obovoid, 0.7–1.5 × 0.4–0.8 cm.

Zambia. N: Chiengi Dist., Lake Mweru, fl. 10–12.x.1949, *Bullock* 1215 (K). W: Mwinilunga Dist., track near Kalene Mission, 6.xi.1962, *Richards* 16900 (K).

Fig. 12.1.**8**. ANCHOMANES ABBREVIATUS. 1, habit (× $^1/_{12}$); 2, ultimate leaf lobes (× $^2/_3$); 3, portion of petiole (× 1); 4, inflorescence with peduncle (× $^1/_{12}$); 5, inflorescence (× 1); 6, spadix (× 1); 7, stamen, side view (× 6); 8, stamen from above (× 6); 9, female flower, side view (× 6); 10, female flower in longitudinal section, showing ovule and placentation (× 6); 11, berry, side view (× 1½); 12, berry in longitudinal section (× 1½); 13, seed, side view (× 1½). 1,3 from *Greenway* s.n., 2 from *Milne-Redhead & Taylor* 7405, 4–10 from *Faulkner* 1149, 11–13 from *Faulkner* 878. Drawn by Ann Webster. From F.T.E.A.

ARACEAE

12,1: **31**

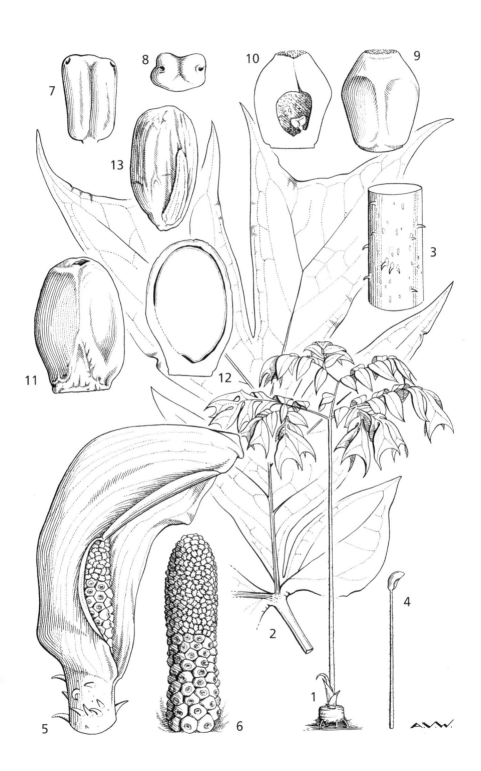

©The Flora Zambesiaca Managing Committee, Kew, 2012

Also in W Africa from Congo to Benin, Burkina Faso, Gambia, Ghana, Guinea-Bissau, Guinea, Ivory Coast, Liberia, Nigeria, Senegal, Sierra Leone, Togo, Central African Republic, Cameroon, Equatorial Guinea, Gabon, Gulf of Guinea Is., Zaire, Chad, Sudan, Uganda, Tanzania and Angola. In riverine and swamp forest, *Brachystegia* and chipya woodland, sometimes on termite mounds; 900–1400 m.

Conservation notes: Restricted distribution in the Flora area, but widespread elsewhere; not threatened.

Flowers are needed to separate these two *Anchomanes* species but are not always present in herbarium specimens. *A. difformis* is predominantly a western species and only enters the Flora area in Zambia, whereas *A. abbreviatus* appears to be confined to Eastern Africa.

A. welwitschii is sometimes considered to be a separate taxon with a sessile discoid stigma and paler colouration. However intermediate forms exist, making separation unreliable. More research on living plants is required.

10. CULCASIA P. Beauv.

Culcasia P. Beauv., Fl. Oware **1**: 3 (1805). —Mayo, Bogner & Boyce, Gen. Araceae: 225 (1997).

Erect or ground-creeping herbs with short clasping roots at and below nodes, branches slender. Leaves numerous, often forming a terminal crown in W African terrestrial species. Petiole sheath persistent, more than half total petiole length, pulvinus inconspicuous; blade obliquely ovate to lanceolate, elliptic or oblong, sometimes falcate, acuminate, base rounded to cuneate, rarely emarginate, rarely pubescent below; resin canals pellucid, linear or punctate, sometimes lacking; primary lateral veins pinnate, often forming a submarginal collective vein, otherwise running into the marginal vein, higher order venation reticulate. Inflorescences 1–12(20) in each floral sympodium, internodes sometimes relatively elongated; peduncle short to relatively long. Spathe erect, usually green to white, boat-shaped or somewhat constricted, convolute basally, gaping apically at anthesis, deciduous to marcescent. Spadix subequal to spathe, subsessile to stipitate, cylindric-clavate; female zone usually densely flowered, shorter than male zone, either contiguous with it or separated by a short zone with sterile male flowers, male zone fertile to apex. Flowers unisexual, perigone absent. Male flowers 2–4-androus, obpyramidal, truncate at apex. Female flowers: ovary 1–3-locular, 1 ovule per locule, placentation basal, stigma sessile, sometimes weakly lobed. Infructescence subglobose to cylindric. Berries subglobose to ellipsoid, 1–3-seeded, glossy, green when young, usually red, sometimes orange to greenish-yellow. Seeds large, ovoid to ellipsoid, smooth, brown.

A genus of about 24 species endemic to tropical Africa, with the centre of diversity in the rainforests of west and central Africa. Both species listed here have been confused in the past with *Culcasia scandens* P. Beauv. (see Letouzey & Ntepé in Taxon **30**: 794, 1981).

Primary lateral veins of leaf long-arcuate, usually running into margin; inflorescences 1–2(3) on each floral sympodium; spathe usually white; female zone of spadix usually sessile or subsessile; in montane forests . . . . . . **1.** *falcifolia*
– Primary lateral veins of leaf usually anastomosing to form looping inframarginal veins; inflorescences 4–8 on each floral sympodium; spathe green; female zone of spadix usually borne on naked basal stipe 0.2–0.7 cm long; in lowland forests . . . . **2.** *orientalis*

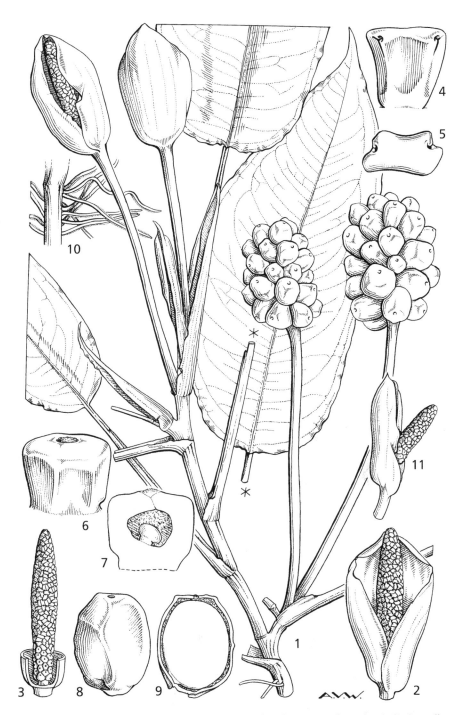

Fig. 12.1.**9**. CULCASIA FALCIFOLIA. 1, habit (× 1); 2, inflorescence, front view (× 1); 3, spadix, spathe removed (× 1); 4, stamen, side view (× 10); 5, stamen, from above (× 10); 6, female flower, oblique view (× 10); 7, female flower in longitudinal section showing ovule and placentation (× 10); 8, berry, side view (× 3); 9, berry in longitudinal section (× 3). CULCASIA ORIENTALIS. 10, roots (× 1); 11, inflorescence, side view (× 1). 1 from *Snowden* 1043, 2–7 from *Dawkins* 361, 8,9 from *Dawkins* 350, 10 from *Faulkner* 162, 11 from *Faulkner* 63. Drawn by Ann Webster. From F.T.E.A.

1. **Culcasia falcifolia** Engl. in Bot. Jahrb. Syst. **26**: 418 (1899). —N.E. Brown in F.T.A. **8**: 175 (1901). —Mayo in F.T.E.A., Araceae: 18 (1985). Type: Tanzania, Uluguru Mts, Nghweme (Nglwenu), 18.x.1894, *Stuhlmann* 8817 (B holotype). FIGURE 12.1.**9**.

Climbing or ground-creeping herb with short clasping roots at and below nodes; internodes 1–14.5 × 0.3–0.4(0.5) cm when dry; epidermis tan-coloured when dry, sometimes densely or sparsely minutely tuberculate. Petiole 6–15 cm long, sheath much more than half total petiole length; blade asymmetric, obliquely ovate to lanceolate, elliptic or oblong, often falcate, 5.5–31 × 3–10.5 cm, 1.9–3.3 times longer than broad, apex acuminate, base rounded to wedge-shaped; resin canals pellucid, usually sparse to almost lacking, sometimes frequent, linear, up to 4 mm long; primary lateral veins long-arcuate, running into margin, usually not forming inframarginal veins. Inflorescence 1–2(3) inserted on a floral sympodium, all except the first subtended by lanceolate bracts; peduncle 4.5–16 cm long. Spathe thickly coriaceous, usually greenish white, white or yellowish, rarely green, yellowish green or orange, obovoid and weakly constricted towards base when closed, 2.5–5 × 0.5–1.4 cm when closed, boat-shaped and gaping to 2 cm wide at anthesis. Spadix subequal to spathe; male zone 2.3–3.2 × 0.5–0.8 cm, subcylindric, narrowing to subacute apex; female zone 0.6–1 × 0.3–0.5 cm, sessile to subsessile, rarely shortly stipitate. Female flowers densely congested, subprismatic; stigma sessile, circular. Mature infructescence ± cylindric with 10–25 berries; berries obovoid to subglobose, to 1.7 × 0.8–1.1 cm, 1(2)-seeded, glossy green when young, becoming orange-red to dark orange-red at maturity. Seed ellipsoid, 0.8–1.2 × 0.6 cm, a dark oval patch on one side.

Zambia. N: Kawambwa Dist., Mbereshi R., 14.ix.1963, *Mutimushi* 445 (K). W: Mwinilunga Dist., banks of Muzera R., 16 km W of Kakoma, 28.ix.1952, *White* 3409 (FHO, K). **Zimbabwe**. E: Mutare Dist., Penhalonga, 24.ix.1950, *Chase* 3217 (K). **Malawi**. N: Chitipa Dist., Mughese Forest Reserve, fl. 28.i.1989, *Thompson & Rawlins* 6247 (K). S: Mt Mulanje, Ruo Gorge, 14.v.1963, *Wild* 6243 (K, LISC). **Mozambique**. Z: Gurué Dist., Gurué Mts, near source of Rio Malema, 5.i.1968, *Torre & Correia* 16952 (LISC). MS: Sussundenga Dist., 20 km W of Dombe, foothills at SE end of Chimanimani Mts, st. 24.iv.1974, *Pope & Müller* 1282 (LISC).

Also in Uganda, Kenya, Tanzania, Sudan, Ethiopia and Congo. Evergreen rainforest, in shade; 600–1600 m.

Conservation notes: Widespread species; not threatened.

2. **Culcasia orientalis** Mayo in F.T.E.A., Araceae: 20 (1985). Type: Tanzania, Lushoto Dist., Usambara Mts, Lunguzu, ii.1962, *Procter* 2009 (K holotype). FIGURE 12.1.**9** and 12.1.**10**.

Climbing herb with short clasping roots at and below nodes; internodes 2.5–5.5 × c.0.5 cm; epidermis tan-coloured when dry, sparsely and distinctly tuberculate. Petiole 6.5–10.5 cm long, sheath usually more than half total petiole length; blade obliquely ovate to elliptic, 8–18 × 3.9–9 cm, ± 2 times longer than broad, shortly acuminate to cuspidate, base rounded or subacute; pellucid glands sparse to frequent, variable in size, punctate or linear, to 0.4 cm long (not seen in *Harder et al.* 2194); sometimes sparse circular excrescences present on lower surface; primary lateral veins forming looping inframarginal veins, at least in upper half of blade. Inflorescence (3)4–8(12), usually 4, inserted on a very short sympodium; all peduncles except the first subtended by lanceolate bracts; peduncle 3–8.5 cm long. Spathe slightly constricted centrally when closed, green, 2–4 × 0.5–0.9 cm when closed, boat-shaped, gaping at anthesis. Spadix subequal to spathe, naked basal stipe 0.2–0.7 × 0.25 cm; female zone 0.5–0.6 cm long; male zone 1.5 × 0.55 cm. Female flowers densely congested, subprismatic; stigma sessile, circular. (Flower characters from F.T.E.A.) Mature inflorescence usually with less than 4 berries; berries obovoid to ellipsoid, 0.8–1.1 × 0.6–0.7 cm when dry, glossy red.

Fig. 12.1.**10**. CULCASIA ORIENTALIS. 1, habit (× ²/₃); 2, inflorescence, side view (× 1½); 3, flowering spadix, spathe removed (× 2); 4, post-floral spadix, spathe removed (× 2); 5, male flower of 3 stamens from above (× 15); 6, female flowers, oblique view (× 15). 1 upper part, 2–6 from *Procter* 2009, 1 lower part from *Faulkner* 1359. Drawn by Ann Webster. From F.T.E.A.

Zambia. N: Kaputa Dist., Sumbu Nat. Park, fr. 5.xii.1993, *Harder et al.* 2194 (K). **Mozambique**. N: Palma Dist., Lake Nhica on R. Rovuma floodplain, Nhica do Rovuma, 25 m, 13.xi.2009, *Goyder* 651 (K, LMA).

Also in Tanzania and Kenya. Riverine, swamp and secondary forest; 10–800 m.

Conservation notes: Restricted distribution in the Flora area, but widespread elsewhere; not threatened.

The species is widespread in the F.T.E.A. area, but only just extends into the Flora Zambesiaca area in N Zambia.

11. **ZANTEDESCHIA** Spreng.

Zantedeschia Spreng., Syst. Veg. **3**: 765 (1826), nom. conserv. —Mayo, Bogner & Boyce, Gen. Araceae: 232 (1997).

Richardia Kunth in Mém. Mus. Hist. Nat. **4**: 437 (1818), invalid name.

Terrestrial, seasonally dormant, sometimes evergreen herbs, tubers depressed globose or with a thick rhizome. Leaves several, basal. Petiole spongiose, with a long sheath. Blade lanceolate, narrowly elliptic, cordate-sagittate or hastate, often variegated; primary lateral veins pinnate, running into a distinct marginal vein, secondary and tertiary laterals parallel-pinnate, higher order venation transverse-reticulate. Inflorescence appearing with leaves. Peduncle erect, longer or equal to petiole. Spathe convolute and obconic basally, upper part spreading widely, not constricted, persistent, white, cream, yellow, pink or purple. Spadix shorter than spathe, sessile or stipitate, male and female zone contiguous, fertile to apex. Flowers unisexual, perigone absent. Male flowers 2–3-androus, stamens free, subsessile, connective truncate at apex, anthers dehiscing by apical pores. Female flowers either a naked gynoecium or rarely (*Z. aethiopica*) surrounded by whorl of c.3 staminodes; ovary ovoid, 1–5-locular, ovules (1)4(8) per locule, anatropous, placentation axillary, stigma small. Berries obovoid or depressed globose, 1 to several-seeded, green, orange or rarely yellow. Seeds ovoid to ellipsoid, strophiolate, testa costate, embryo axile, elongate, endosperm copious.

A genus of 8 species with most occurring in South Africa. Two species in the Flora area, with one other introduced.

Introduced species.

Zantedeschia aethiopica (L.) Spreng., White Arum.

Robust evergreen plant up to 60 cm high. Petiole smooth. Inflorescence without a dark purple area at base of spathe inside and on limb, which is consistently ivory-white, broad and widely spreading backwards. Typically the spadix is bright yellow; female flowers interspersed with staminodes.

Zimbabwe. C: cultivated, *Biegel* 4558 (SRGH). **Mozambique**. M: recorded in Sabonet checklist.

Cultivated throughout the world, now naturalised in various parts of tropical America, in S Europe, the Philippines, New Zealand and elsewhere. It is easily distinguished from *Z. albomaculata* by its pure white spathes and the numerous clavate-spathulate staminodes scattered among the female flowers.

1. Leaves hastate to cordate at base 2
- Leaves narrowly lanceolate, cuneate at base . . . . **2.** *rehmannii*
2. Spathe pure white, numerous clavate-spathulate staminodes scattered among female zone . . . . *aethiopica*
- Spathe cream-yellow to green-yellow, female zone lacking staminodes . . . .
. . . . **1.** *albomaculata*

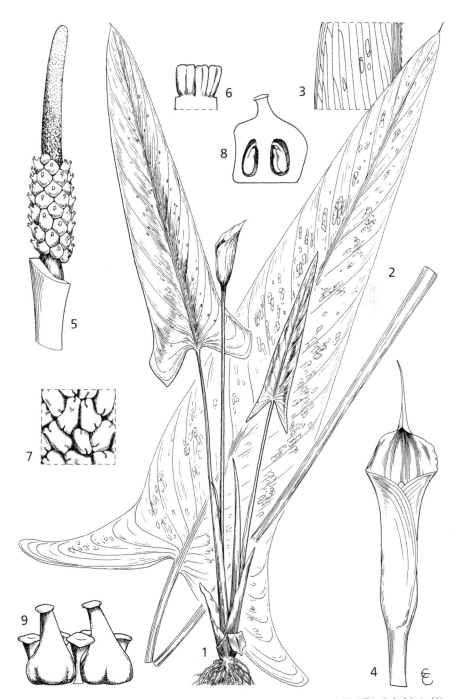

Fig. 12.1.**11**. ZANTEDESCHIA ALBOMACULATA subsp. ALBOMACULATA. 1, habit (× ¼);
2, leaf (× ½); 3, detail on leaf showing marginal venation (× 1); 4, inflorescence (× ²/₃); 5,
spadix, spathe removed (× 1½); 6, stamens, side view (× 6); 7, stamens from above (× 10); 8,
female flower in longitudinal section showing ovules and placentation (× 6). ZANTEDESCHIA
AETHIOPICA. 9, ovaries and staminodes (× 6). 1 from *Milne-Redhead* 3823 and *Richards* 10250, 2
from *Richards* 13859, 3 from *Richards* 10250, 4–8 from *Milne-Redhead* 3823, 9 from *Nicholson* 4263.
Drawn by Eleanor Catherine. From F.T.E.A.

1. **Zantedeschia albomaculata** (Hook.) Baill. in Bull. Mens. Soc. Linn. Paris **1**: 254 (1880). —Letty in Bothalia **11**: 17 (1973). —Mayo in F.T.E.A., Araceae: 36 (1985). —Cook, Aq. Wetl. Pl. Sthn. Africa: 61 (2004). Type: South Africa, KwaZulu-Natal, plant cultivated at Kew, received from Messrs. Backhouse of York in 1859 (K holotype).

Richardia albomaculata Hook. in Bot. Mag. **85**: t.5140 (1859).

Arodes albomaculatum (Hook.) Kuntze, Revis. Gen. Pl. **2**: 740 (1891).

Perennial herb. Tuber depressed globose, 3–4 cm wide or more. Petiole 23–57 cm long, sheath ± half petiole length. Leaf blade 19–47 × 6–24 cm, either conspicuously white-maculate or immaculate, hastate-sagittate, anterior lobe triangular-lanceolate to ovate, posterior lobes to 12 cm long, rounded to acute, broad to narrow, directed downward to widely spreading. Peduncle 24–58 cm long. Spathe 5.5–15 × 7.5–13 cm when flattened, cream-yellow to green-yellow both internally and externally, internally usually with a dark maroon patch at base; spathe tube obconic; spathe limb apiculate, somewhat recurved. Spadix 2.8–5 cm long, stipitate, stipe 0.3–0.4 cm; female zone 0.8–1.8 × 0.5 cm, lacking staminodes; male zone 1.3–3 × 0.3–0.4 cm. Female flowers greenish; ovary subglobose, 2–3 mm wide, narrowing into a c.1 mm slender style, stigma capitate. Male flowers yellow.

Subsp. **albomaculata** FIGURE 12.1.**11**.

Richardia hastata Hook. in Bot. Mag. **86**: t.5176 (1860). Type: South Africa, Natal, plant cultivated at Kew, received from Messrs. Veitch of Exeter, n.d. (K holotype).

Richardia angustiloba Schott in J. Bot. **3**: 35 (1865). Type: Angola, Pungo Andongo, iii.1857, *Welwitsch* 230 (BM holotype, K).

Zantedeschia hastata (Hook.) Engl. in Bot. Jahrb. Syst. **4**: 64 (1883).

Zantedeschia angustiloba (Schott) Engl. in Bot. Jahrb. Syst. **4**: 64 (1883). —Hepper in Kew Bull. **21**: 493 (1968); in F.W.T.A., ed.2 **3**: 120 (1968).

Zantedeschia tropicalis (N.E. Br.) Letty in Bothalia **7**: 456 (1961). Type: Malawi, Namadzi, 1897, *Cameron* s.n. (K lectotype), lectotypified by Letty (1961); Zimbabwe, Harare, 6 mile spruit, *E.Cecil* 149 (K syntype).

Zantedeschia oculata (Lindl.) Engl., Pflanzenr. **73** IV 23d: 68 (1915). Type as for *Richardia hastata*.

Zambia. N: Mbala Dist., upper part of Nkali dambo, fl. 5.i.1952, *Richards* 342 (K). W: Chingola Dist., Chingola, fl. 7.xi.1968, *Mutimushi* 2809 (K). **Zimbabwe**. C: Harare, fl. 28.i.1968, *Plowes* 2675 (PRE, SRGH). E: Mutare Dist., Engwa, fl. 1.ii.1955, *Exell* 4 (LISC). **Malawi**. N: Nkata Bay Dist., S Viphya forest plantation, fl. 6.ii.1992, *Goyder et al.* 3625 (K, PRE). C: Dedza Dist., Mt Dedza forest, fl. 9.ii.1968, *Salubeni* 956 (K, PRE). S: Blantyre Dist., Namadzi (Namasi), 1897, *Cameron* s.n. (K).

Also from Tanzania, Congo, Angola, Lesotho, Swaziland, South Africa (Eastern Cape to KwaZulu-Natal, Orange Free State and former Transvaal); possibly also in Nigeria and Cameroon. Usually found in marshy ground near rivers, but also among rocks on grassy hillsides and forest margins; 1100–1800 m.

Conservation notes: Widespread taxon; not threatened.

Subspecies *macrocarpa* Engl. occurs in South Africa and differs from subsp. *albomaculata* in having triangular-hastate, sparsely speckled leaves, straw-coloured spathes truncate at the apex, and a few large berries up to 2 cm in diameter.

2. **Zantedeschia rehmannii** Engl. in Bot. Jahrb. Syst. **4**: 63 (1883). —Letty in Bothalia **11**: 11 (1973). Type: South Africa, New Castle, 1875, *Rehmann* 80 (Z holotype, K).

Richardia rehmannii (Engl.) Krelage in Gard. Chron., n.s. **1893**(2): 564 (1893).

Perennial herb. Tuber unknown. Petiole 10–20 cm long, sheathing for ± half its length. Leaf blade 15–40 × 2–7 cm, lanceolate, base gradually tapering to petiole, apex acuminate, usually immaculate. Peduncle up to 60 cm long. Spathe 9–12 cm long, white to pink, with no internal dark patch, rolled into a narrow funnel; spathe limb spreading slightly, tapering at apex. Spadix usually stipitate, 3.5–5 cm long; female zone 0.7–1 cm long; male zone 1.6–2.2 cm long.

Mozambique. [no locality given], st. n.d., ?*Foster* s.n. (K).
Also in South Africa. Uncertain locality and habitat; in South Africa found between rocks on grassy hillsides.
Conservation notes: Uncertain locality in the Flora area; possibly Vulnerable.
In the Flora area known only from one sterile specimen, the identification of which is confirmed by the distinctive leaf shape. As no detailed locality is given more field work is required to determine its distribution. The description here is prepared from South African material.

12. AMORPHOPHALLUS Decne.

Amorphophallus Decne. in Nouv. Ann. Mus. Hist. Nat. **3**: 366 (1834), conserved name
—Mayo, Bogner & Boyce, Gen. Araceae: 235–239 (1997). —Ittenbach in Englera **25**: 1–263 (2003).
Hydrosme Schott, Gen. Aroid.: t.33 (1858).

Seasonally dormant herbs (rarely evergreen), often large, stem usually a depressed-globose tuber. Leaf usually solitary in adult plants (rarely 2–3), sometimes 2–3 in seedlings. Petiole long, usually smooth, erect, cylindric, sometimes very thick, not pulvinate apically, usually spotted and marked in a variety of patterns, sheath very short. Blade trisect, primary divisions pinnatisect, bipinnatisect or dichotomously further divided, tubercles rarely present at junction of divisions, secondary and tertiary divisions ± regularly pinnatifid to pinnatisect, ultimate lobes oblong-elliptic to linear, acuminate, decurrent; primary lateral veins of ultimate lobes pinnate, forming a distinct submarginal collective vein, higher order venation reticulate. Inflorescence always solitary, preceded by cataphylls, usually flowering without leaves, rarely with leaves. Peduncle very short to long, similar to petiole. Spathe variously coloured, marcescent and finally deciduous, usually clearly differentiated into tube and limb, sometimes constricted between; tube convolute (rarely connate), campanulate to cylindric or ventricose, inner surface smooth, longitudinally ribbed, verruculose near base, scabrate or densely covered with scale- or hair-like processes; limb erect to spreading, smooth, ribbed or variously undulate or frilled at margins. Spadix shorter or much longer than spathe; female zone shorter, equal or longer than male zone; male zone cylindric, ellipsoid, conoid or obconoid, usually contiguous with female, sometimes separated by a sterile zone which may be naked or bear prismatic subglobose or hair-like sterile flowers; terminal appendix present (rarely absent), usually erect, rarely horizontal or pendent, very variable in shape, usually ± conoid or cylindric, sometimes ± stipitate or basally narrowed, usually smooth or bearing staminode-like structures near base or entirely covered with staminodes, sometimes corrugate or densely to sparsely hirsute, or grossly and irregularly crumpled. Flowers unisexual, perigone absent. Male flowers 1–6-androus, stamens free or sometimes connate in basal flowers or throughout male zone, short, filaments absent or distinct, connective fairly thick, sometimes projecting beyond thecae, thecae obovoid or oblong, opposite, dehiscing apically by an apical (rarely lateral) pore or transverse slit. Female flowers usually crowded, sometimes ± distant, ovary ± ovoid, 1–4-locular, ovules 1 per locule, anatropous, funicle very short to distinct, erect, placenta axile to basal, style absent to very long, stigma variously shaped, sometimes brightly coloured. Infructescence ± cylindric; berries 1 to few-seeded, orange to red, rarely blue or white. Seeds ellipsoid, smooth.

A paleotropical genus of 196 species, the majority in Asia and Malesia. Africa has 35 native species.
The Flora area contains several recently-described species of *Amorphophallus*, including some narrow endemics. It is suspected that with further studies, more such species will be found.
The genus *Amorphophallus* presents several problems for identification. The first is

that the leaf and inflorescence usually appear separately. Identification from the leaf alone is very difficult, especially in dried material where it loses some of the distinctive markings. The second is that high quality collections of the inflorescence are rare and without these confident identification is problematic. The best way to preserve them is in alcohol.

1. Spadix ± shorter or equal to spathe 2
 - Spadix distinctly longer than spathe 4
2. Spadix distinctly shorter than spathe, basal inner side of tube with ± parallel running ribs, evenly high and not interrupted in lateral view . . . . . **1.** *abyssinicus*
 - Spadix usually ± as long or a little longer than spathe, basal inner side of spathe with platform-like and wart-like emergences, forming ribs which are unevenly high in lateral view and interrupted 3
3. Smooth pollen exine, appendix often shorter than spathe; ovary unilocular, ratio of female zone to male zone 0.5–0.7:1 . . . . **2.** *mossambicensis*
 - Striate pollen exine, appendix longer than spathe; ovary bi- to trilocular, ratio of female to male zone 0.35–0.4:1. . . . .**3.** *richardsiae*
4. Peduncle clearly elongated, 30–75 cm long. . . . . **6.** *maximus*
 - Peduncle short, to 12 cm long 5
5. Tube asymmetric-ovoid; basal inner side of limb with short warts, papillae or hair-like (0.2–1 mm long). . . . . **4.** *goetzei*
 - Tube strongly compressed; basal inner side of limb with small, tongue-shaped colourless or purple or red-brown emergences 0.3–2 × 0.5 mm, reduced in size higher up so inside surface rough . . . .**5.** *impressus*

1. **Amorphophallus abyssinicus** (A. Rich.) N.E. Br. in F.T.A. **8**: 160 (1901). —Mayo in F.T.E.A., Araceae: 29 (1985). Type: Ethiopia, Tigre, Tekeze (Takazze) R. valley, near Djeladjeranne, *Quartin Dillon* s.n. (P holotype).

  Arum abyssinicum A. Rich., Tent. Fl. Abyss. **2**: 352 (1850).

Subsp. **unyikae** (Engl. & Gehrm.) Ittenb. ex Govaerts & Frodin, World Checklist Araceae: 81 (2002). Type: Tanzania, Mbeya Dist., Unyika, Dorf Toola, 8.xi.1899, *Goetze* 1413 (B holotype, K photo).

  Amorphophallus unyikae Engl. & Gehrm. in Engler, Pflanzenr. **73** IV 23c: 72 (1911).

Tuber discoid to sub-globose, with short to elongate, cylindrical daughter tuber. Leaf emerging after inflorescence, 40–100 cm high; petiole 20–80, reddish to greenish, with darker spots and whitish-grey wax cover; blade 20–60 cm wide. Cataphylls two, 6–41 cm long, pale green to pink, with dark green-purple markings, inner one longer than peduncle. Inflorescence solitary, very rarely two emerging from one tuber. Peduncle 7–19 cm long, somewhat glossy, background dark green to brownish, with elongate brownish markings. Spathe 7–18 cm long, divided by strong constriction into a cylindrical to conic-ovoid tube and a widely triangular to roundish open blade. Tube 3–7 × 2–4 cm, internally with a pale, green strip immediately below constriction; outside greenish, purple to brownish, with dark green-brown speckling or striations; basal inner side dark purple to red brown, with forking to parallel, congested, narrow ribs. Limb open, erect to tilted, (4)6–12 cm long, triangularly tapering to broadly elliptic, inside dark to light purple, outside greenish to purple; margin smooth or undulate. Spadix sessile, 6.5–12 cm long, (1.5)2–5(9) cm shorter than spathe. Female zone cylindrical, 1–2 cm long, flowers congested. Male zone cylindrical, 2–3.5 cm long, flowers very densely congested. Appendix ± cylindrical to short conical, 4–10 cm long, purple red-brown, finely wrinkled, widening in first quarter, gradually tapering from second quarter and ending in a rounded tip, inside ± solid. Staminodes absent; sterile zone between female and male zone absent; ratio of female to male zone 0.4–0.5:1. Female flowers 3 mm long; ovary globose, egg-shaped, 2 × 2 mm, pale green; stigma sessile, 1.5–2 mm wide, reddish purple to brown, flattened globose, smooth to slightly papillose. Male flowers 2–3 × 1.5–2 mm; anthers free, 1–1.5 × 1.5 mm, ovoid, light brown to beige; filaments free or basally fused, short or 1–2 mm long; pores apical, connective channelled, purple. Infructescence solitary or with leaf; stalk 2–4 times longer than during flowering period; cylindrical, c.8 × 3 cm, often with remains of spathe. Berries 0.8–1 × 0.6–0.8 cm, 1–2(3)-seeded, ovoid, red. Seeds ovoid to flattened-ovoid, dark brown to black, 0.5–0.8 × 0.3–0.4 cm.

Caprivi: Impalila Is., central portion, fl. ii.1976, *du Preez* 30 (PRE). **Zambia**. B: Kalabo Dist., 3.2 km W of Kalabo, fr. 12.xi.1959, *Drummond & Cookson* 6446 (K). W: Mwinilunga Dist., slope E of Mantonchi Farm, fl. 17.x.1937, *Milne-Redhead* 2818 (BR, K). S: Mazabuka Dist., Central Research Station, Mazabuka, 13.xi.1931, *Trapnell* 516 (K). **Zimbabwe**. N: Gokwe Dist., Gokwe R., fl. 20.xi.1963, *Bingham* 970 (K). W: Hwange Dist., near Victoria Falls, fl. xi.1955, *Levy* s.n. (PRE). C: Harare (Salisbury), Enterprise Road, fl. 13.xi.1951, *Greatrex* 34757 (K). **Malawi**. N: Nkhata Bay Dist., Chinteche, fr. 30.xii.1978, *Phillips* 4503 (K, MO).

Also in Congo. In *Brachystegia* and mopane woodland and grassland, on sand or red soils, often on termite mounds; 500–1700 m. Flowers in October to December.

Conservation notes: Widespread; not threatened.

Roots reported to be edible in Zambia (*Mitchell* 24/36).

Amorphophallus abyssinicus is the best known and most widely distributed species in Africa. Its morphological variability and wide distribution led to a large number of synonyms (e.g. Redhead 1950, Mayo 1985). Since then, Ittenbach has described *A. richardsiae* as a new species, reinstated *A. mossambicensis* as an accepted name, and included 3 subspecies within *A. abyssinicus*. Subsp. *unyikae* marks the southern limit of the distribution range. Subsp. *abyssinicus* is more common throughout the species' range, extending west to Togo and north to Sudan and Ethiopia. It has also been recorded as possibly occurring as far south as Malawi, but no specimens have been seen. A third subspecies (subsp. *akeassii* Ittenb.) is found in West Africa.

The different subspecies are difficult to differentiate in herbarium specimens, especially if sterile or fruiting. Subsp. *unyikae* is differentiated by the lack of a style and having the basal inside of spathe with irregular low, forking to parallel, congested, narrow ribs and the appendix basally hardly or not at all constricted.

2. **Amorphophallus mossambicensis** (Schott) N.E. Br. in F.T.A. **8**: 150 (1901). Type: Mozambique, Tambara Dist., Lupata Mts, *Peters* s.n (B† holotype). Lectotype: illustration t.33 in Schott, Gen. Aroid. (1858), lectotypified here.

> *Hydrosme mossambicensis* Schott in Gen. Aroid.: t.33 (1858).
> *Corynophallus mossambicensis* (Schott) Kuntze, Revis. Gen. Pl. **2**: 741 (1891).
> *Amorphophallus swynnertonii* Rendle in J. Linn. Soc., Bot. **40**: 219 (1911). Type: Mozambique, Gazaland, Madanda, 1906, *Swynnerton* 717a (BM holotype).

Tuber flattened-globose, 4–8 × 2–4 cm, roots emerging on upper surface. Leaf 40–70 cm high; petiole 25–40 cm long; blade 30–60 cm wide; rachis broadly winged, terminal leaflets broadly elliptic to elongate, 7–10 × 2–3 cm. Cataphylls 2?, 8–20 × 2–3 cm, outside red brown with whitish spots, inside dark green. Inflorescence emerging before leaf, to 60 cm high. Peduncle 12–28 cm long. Spathe 15–28 cm long, arranged in a ± cylindrical tube and a broadly elliptic to long, triangular open limb. Tube 4–8 × 2–4 cm; basally rounded, outside probably green-purple to purple, with darker and small spots; inside ribbed basally, formed of narrow platform-like and irregularly shaped emergences, 1.2 mm high, appearing unevenly high (like mountain chains) from the side, not running parallel, ribs very congested, red-brown, a light ring above this present at level of the constriction. Limb erect, 10–20 cm long, elongate or elongate-triangular, greenish purple to purple both inside and out, becoming dark purple toward margin; margin undulate to frilly lobed. Spadix sessile, 19–28 cm long, ± as long as spathe. Female zone cylindrical to conical, 2–2.5 × 1–1.6 cm, flowers congested. Male zone cylindrical to obconic, 2.5–5 × 1–2 cm, flowers congested. Appendix 13–18 × 1–1.5 cm at base, narrowly conical, lower third sometimes with very small warts, loosely and unevenly dispersed; basally slightly constricted, red-brown to purple. Staminodes sometimes as very small irregular warts in lower third of appendix. Sterile zone between female and male zone absent; ratio of female to male zone 0.5–0.7:1. Female flowers 3–4 mm long; ovary elongate to globose, 2 × 2–3 mm, uni(bi)locular; stigma sessile, circular from above, broadly elliptic to circular from side, unlobed or rarely with two flat-globose elevations. Male flowers: 2–4 stamens forming one flower; anthers free, elongate, rectangular to cylindrical; filament 0.5–2 mm long, free; opening by apical pores. Infructescence unknown.

Zambia. S: Choma Dist., Muckle Neuk, 12 km N of Choma, fl.& lf. 19.xi.1954, *Robinson* 956 (K). **Zimbabwe.** N: Mazoe Dist., Mazoe, fl. xii.1917, *Eyles* 883 (BM, K). **Mozambique.** MS: "Sofala", coast (near Beira?), fl.& lf. 31.x.1906, *Johnson* 23 (K).

Also known from Tanzania and Congo. Often dispersed in open forest and miombo woodland in sandy loam soils; 100–1350 m. Flowers from October to December.

Conservation notes: Widespread; not threatened.

A. mossambicensis and its synonyms in the past have been put with *A. abyssinicus.* However *A. mossambicensis* can be distinguished from *A. abyssinicus* by the interrupted irregular ribs of the basal region inside the spathe tube. Other differing characteristics include pollen morphology.

The habit drawing (fig. 1 of t.33) in Schott (1858) is a copy of a drawing by Klotzsch (published later in 1864 by Peters, Naturwiss. Reise Mossamb.) to which Schott had prior access ("a Klotzschio data"). Under Art. 44.1 of the Vienna Code (2006), the name *Hydrosome mossambicensis* was validly published by Schott. All sources make it clear that there was only one original element on which this name was based.

3. **Amorphophallus richardsiae** Ittenb. in Willdenowia **27**:157 (1997). Type: Zambia, Mbala Dist., Kalambo Falls, 15.xi.1960, *Richards* 13579 (K holotype).

Tuber unknown. Leaf unknown. Cataphylls 2?, 2.3–13.5 × 1.2–6 cm, cream coloured with a hint of rose and elongate-broadly elliptic brownish spots. Inflorescence to 40 cm high, emerging before leaf. Peduncle 12–15 cm; greenish with overlapping, elongate to broadly elliptic, small, dark green to brown spots. Spathe 16–24 cm long, divided by a distinct constriction into an asymmetric, ovoid to triangular tube and a wide, triangular to broadly elliptic, open limb. Tube 5–8 × 4–6 cm, the basal part externally brownish green, becoming greenish towards constriction; veins brownish to lilac, densely covered with dark green to brownish circular spots; inside basally purple with strongly raised veins, these crowded with warts and flag-shaped emergences, veins often appearing as even or unevenly high ribs; a greyish green, 1–2 cm wide waxy band inside the constriction. Limb open, 10–17 cm long, erect-oblique, outside dirty green with lilac veins, inside brown-purple, margins undulate. Spadix sessile, 21–27 cm long, equal or up to 6 cm longer than spathe. Female zone cylindrical, 1.2–2.5 × 1.2–1.5 cm, flowers congested. Male zone cylindrical to slightly ovoid, 3.2–6 × 1–1.5 cm, flowers congested. Appendix 14–20 cm long, elongate conical to cylindrical, basally somewhat constricted and furrowed, purple. Staminodes absent; sterile zone between female and male zone absent. Ratio of female to male zone 0.35–0.4:1. Female flowers 2.5–3.5 mm long; ovary globose-ovoid, 2–2.5 × 1.5–2 mm, 2–3-locular, dark yellow; stigma sessile, 2–3-lobed, 1–2 × 1–1.5 mm, circular to oval in cross section, brown-black. Male flowers with anthers cubic, free, light brown; filament 0.7–1 × 0.7–1 mm, free; connective slightly grooved with a dark stripe; pores apical. Infructescence unknown.

Zambia. N: Mbala Dist., Mbala (Abercorn), by PWD Camp, 1500 m, fl. 14.xi.1959, *Richards* 11777 (K).

Endemic to Zambia near the Tanzania border, from Mbala to Kalambo waterfall. In grassland and bushland, on red soils; 1200–1500 m. Flowers from November to December.

Conservation notes: A narrow endemic, known from only 3 specimens; probably Vulnerable D2.

Although very closely related, *A. richardsiae* is different from *A. mossambicensis* in many morphological characteristics. It differs mainly in the stripey pollen exine, an appendix which is longer than the spathe, the bi- to trilocular ovary, the stigma and in the ratio of female to male zone (0.35–0.4:1).

4. **Amorphophallus goetzei** (Engl.) N.E. Br. in F.T.A. **8**: 150 (1901). —Mayo in F.T.E.A., Araceae: 31 (1985). Type: Tanzania, Kilodsa Dist., E slope of Vidunda Mts, 31.xii.1898, *Goetze* 407 (B holotype).

Hydrosme goetzei Engl. in Bot. Jarhb. Syst. **28**: 355 (1900).
Amorphophallus anguineus Peter in Nachr. Ges. Wiss. Göttingen, Math.-Phys. Kl. **1929**(3): 192, 195, 215, figs.1, 2 (1930). Types: Tanzania, Kilosa Dist., near Kimamba, *Peter* 32458 (B syntype); Mukondokwa valley, *Peter* 32615 (B† syntype).

Tuber flattened-globose, 4–13 × 2.5–7 cm, probably developing much bigger, apparently not forming offsets. Leaf 60–100 cm high; petiole 40–60 cm long; blade 40–60 cm wide; rachis winged only from 2nd order branching onwards; terminal leaflets 5–15 × 1–8 cm. Cataphylls 3, broadly lanceolate, inner one longer than peduncle, 10–23 cm, spotted, membranous. Inflorescence 25–100 cm high, appearing before the leaf. Peduncle 5–20 cm long, light-green to green with small brownish spots. Spathe 10–38 cm long, divided by a strong constriction into an asymmetric, ovoid tube and a triangular, acute open limb. Tube 4–12 × 3–8 cm, externally whitish green with dark green, irregularly shaped spots; inside dark purple basally with short 0.2–1 mm warts and papillae, often fused basally into irregularly shaped longitudinal ribs which break up into distinct papillae closer to the constriction. Limb open, erect to oblique, 6–12 cm long, elongate-triangular, margins undulate, outside dark green to light purple, inside purplish. Spadix sessile or stipitate for 6 mm at most, 15–80 cm long, 1.5–2.5 times as long as spathe, emerging at an angle from peduncle. Female zone cylindrical to conical, 1–3 × 0.7–1.3 cm, flowers not always congested. Male zone obconical to ovoid, 1.5–5 × 1–3.5 cm, flowers congested. Appendix 11–72 cm long, elongate club-shaped to conical, bulging in lower ¼ (3–6 cm wide) and then evenly attenuating; usually distinctly basally constricted and with longitudinal furrows, apically acute, light purple with a waxy grey-green flush. Staminodes absent; sterile zone between female and male zone absent or up to 3 mm; ratio of female to male zone 0.5–1.1:1. Female flowers 2–3 mm long; ovary 2 × 1.5 mm, elongate ovoid to globose or somewhat flattened, 1–2-locular, with one basal ovule-attachment, green. Stigma subsessile 1 × 1.5–2 mm, unlobed or somewhat 2-lobed, brownish to beige. Male flowers with anthers elongate to ovoid, yellow; filaments not visible or less than 0.1–0.4 mm long, free or partially fused at base; pores apical, 1(2) circular pores per theca; connective channelled, dark purple. Infructescence stalk slightly elongated, c.20 cm long; infructescence cylindrical, 8–16 × 2–3.5 cm with remains of spathe. Berries ovoid, 0.7–1.4 × 0.5–0.9 cm, 1–2-seeded; seeds elongate ovoid, 6–9 ×4–6 mm, rough.

Mozambique. Z: Lugela Dist., Namagoa, lf.,fl.& fr. c.1944, *Faulkner* 345 (BR, K, PRE). Also known from Tanzania and Congo. In evergreen forests and river valleys; c.200 m. Flowers from November till January.
Conservation notes: Limited distribution in Flora area; possibly Near Threatened. The tuber is said to be very poisonous (*Wallace* 570 from Tanzania).
According to Ittenbach (2003) this species is smaller in the Flora area, the southern and western limits of its range. In F.T.E.A. Mayo mixed and described two species under this name. In 1997, Ittenbach described a new species (*A. impressus*) citing a specimen that Mayo treated under *A. goetzei* as the type (see below).

5. **Amorphophallus impressus** Ittenb. in Willdenowia **27**: 154 (1997). Type: Tanzania, Songea Dist., S of Lumecha bridge, 4.v.1956, *Milne-Redhead & Taylor* 9998 (K holotype). FIGURE 12.1.**12**.

Amorphophallus goetzei sensu Mayo in F.T.E.A. (1985), in part for *Milne-Redhead & Taylor* 9998 and others.

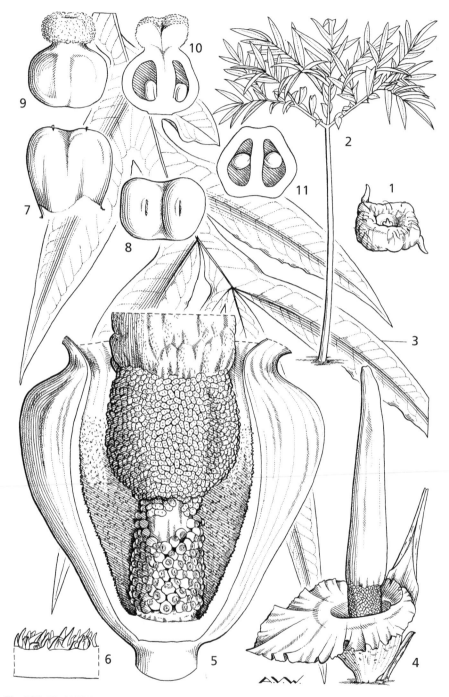

Fig. 12.1.**12**. AMORPHOPHALLUS IMPRESSUS. 1, tuber ($\times \frac{1}{6}$); 2, leaf ($\times \frac{1}{12}$); 3, ultimate leaf lobes ($\times \frac{2}{3}$); 4, flowering plant ($\times \frac{1}{3}$); 5, lower part of spadix, part of spathe removed ($\times 1$); 6, inner surface of spathe-tube towards base, transverse section ($\times 4$); 7, stamen, side view ($\times 10$); 8, stamen from above ($\times 10$); 9, female flower, side view ($\times 6$); 10, female flower in longitudinal section showing ovules and placentation ($\times 6$); 11, female flower in transverse section ($\times 6$). All from *Milne-Readhead & Taylor* 9998. Drawn by Ann Webster. From F.T.E.A.

Tuber flattened, 13 × 7 cm deep. Leaf solitary, c.120 cm high; petiole c.100 cm long, green with purple spots; blade green, c.100 cm wide, rachis with broad-triangular to oval wings; leaflets elongate-oval, acuminate, margin slightly undulate or dentate. Cataphylls whitish to pink and with numerous greenish to brownish oval to roundish spots. Inflorescence ± at ground level, appearing before leaf; peduncle short, 5–10 cm long, brown-green with or without green or olive spots, becoming purplish towards spathe. Spathe 12–25 cm long, separated by a strong constriction into a depressed, urceolate tube and an open triangular limb; tube 3.5–10 × 4–9 cm, outside creamy to flesh-coloured, lower part greyish brown to green, with dark green veins and small green to brown spots; inside pinkish, purple to reddish brown, lower part with small, tongue-shaped colourless or purple or red-brown emergences of 0.3–1 × 0.5 mm, reduced in size higher up causing inside surface to be rough. Limb spreading, strongly undulate, elongate-triangular to 15 cm long, outside dark green to brownish purple, with thick purple veins, inside pink to purple, margin maroon. Spadix sessile or very shortly stipitate, 18–40 cm long, longer than spathe; female zone cylindrical, 1–3 × 0.8–1.2 cm, flowers congested or somewhat distant; male zone obconical to ovoid, narrowed at base, 1.5–4 × 1–3 cm, flowers congested; sterile zone between female and male zone 1–10 mm long; appendix 20–35 cm long, thick-conical to slightly clavate, slightly constricted at base, 2–4 cm in diameter at widest point, dark purple, verruculose; staminodes absent; ratio of female to male zone length 0.4–0.6:1. Female flowers 2–4 mm long; ovaries globose-ovoid, 1.5–2 × 1.5–2 mm, yellowish green or bright green, (1)2(3)-locular; stigma sessile, dark yellow, papillose, with 2(3) small roundish humps or rarely roundish and smooth. Male flowers with anthers free, elongate-ovoid to cubic, creamy to yellowish orange. Infructescence unknown.

Malawi. S: Lake Malawi, Monkey Bay, fl. 22.xii.1983, *Gassner* 244 (K spirit); Nyasaland (Njassaland), Karenga (Karanga), fl. 22.xii.1952, *Willamsen* 133 (BM).

Also known from Tanzania. In dry areas on rocks and riverine forest; c.500 m.

Conservation notes: Limited distribution in Flora area; possibly Near Threatened.

Specimens of this species were previously identified as *A. goetzei* before being recognized as new by Ittenbach (1997).

6. **Amorphophallus maximus** (Engl.) N.E. Br. in F.T.A. **8**: 157 (1901). —Mayo in F.T.E.A., Araceae: 32 (1985). Type: Kenya, near Mombassa, *Hildebrandt* 2018 (B† holotype).

Hydrosme maxima Engl. in de Candolle & de Candolle, Monogr. Phan. **2**: 323 (1879).

Subsp. **fischeri** (Engl.) Ittenb. ex Govaerts & Frodin, World Checklist Araceae: 90 (2002). Type: Tanzania, Mwanza Dist., Usukuma, watershed of Simiyu R., n.d., *Fischer* 618 (B holotype).

Hydrosme fischeri Engl. in Bot. Jarhb. Syst. **15**: 460 (1893).
Amorphophallus fischeri (Engl.) N.E. Br. in F.T.A. **8**: 158 (1901).
Amorphophallus schliebenii Mildbr. in Notizbl. Bot. Gart. Berlin-Dahlen **13**: 410 (1936). Type: Tanzania, Lindi Dist., Lake Lutamba, 10.xii.1934, *Schlieben* 5707 (B holotype, BM, BR)

Plant to 2 m. Tuber flattened, irregularly globose, 3 × 9 cm, white to cream. Leaf appearing after inflorescence, solitary; petiole 25–100 × 1–4 cm; blade 40–120 cm wide; terminal leaflets elongate 10–19 × 1.5–2.5 cm. Cataphylls 2–3, 6–23 cm long, silvery grey with purplish markings. Inflorescence emerging with leaf shoots already formed below ground; smelling of rotting carcass and urine. Peduncle 20–130 cm, dark greyish green, with numerous small, elongate markings, glaucous. Spathe 18–23 × 10–22 cm, divided by strong constriction into a wide globose tube and a ± ovoid to triangular, open limb. Tube 5–8 × 4–6 cm, globose to flattened globose, somewhat broader than high; dark purplish brown outside at constriction, with darker veins, glaucous; inside basally chestnut brown to purple, with irregularly shaped warts. Limb open, 12–17 × 8–22 cm, erect, pressed against appendix on one side during anthesis; outside chestnut brown to purple, with raised veins; inside chestnut brown to purple; margin undulate. Spadix stipitate for 1–3 mm, ± twice as long as spathe, 20–51 cm long; female zone conical, 1–3 × 1–2.4 cm, flowers congested to somewhat distant; male zone obconic, 1.7–4 × 1.5–3 cm, flowers congested. Appendix long conical, 20–44 × 1.8–3 cm, purple, very rough, slightly constricted at base and often deeply furrowed; with numerous, clear drops, crystallizing rapidly during pistillate phase, hollow inside. Staminodes few, only basally as angular, small elevations; sterile zone between female and male zone 1–5 mm wide; ratio of female to male zone 0.5–0.8:1. Female flowers 3–4 mm long; ovary somewhat compressed-globose; 2–3-locular, one basal ovule-attachment per locule, pale green, apically dark brown; style 1.5–2 mm, dark brown; stigma somewhat broader than ovary, 2–3-lobed, verrucose, brownish to beige. Male flowers with anthers free, 1.3–1.5(2) mm, dark yellow; pores apical. Infructescence with slightly elongated peduncle, cylindrical, 14 × 4 cm. Berries ovoid, 0.5–0.9 cm, when dry.

Zimbabwe. N: Makonde Dist., Lomagundi, fl. x.1920, *Eyles* 2699 (K, PRE). C: Harare (Salisbury), fl.& lf. 28.xi.1926, *Eyles* 4559 (K). **Mozambique**. N: Nacala Dist., 15 km from Nacala Nova (Maaia) towards Nacala Velha, fl. 3.xii.1963, *Torre & Paiva* 9385 (BR, LISC, WAG). T: Moatize Dist., near Boroma, Zambesi, fl. i.1890–1891, *Menyharth* 922 (K).

Also in Tanzania. Often close to river and lake margins, on dark loam soils; 10–1500 m. Flowers from October to January.

Conservation notes: Fairly widespread taxon; not threatened.

Easily recognisable as the tallest *Amorphophallus* inflorescence in the region.

13. PISTIA L.

Pistia L., Sp. Pl.: 963 (1753); Gen. Pl., ed.5: 411 (1754). —Engler, Pflanzenr. IV **23f**: 258 (1920). —Mayo, Bogner & Boyce, Gen. Araceae: 286 (1997).

Small, free-floating evergreen herbs with pendent feathery roots; stem very short, acaulescent, stoloniferous. Leaves several in a rosette, densely pubescent, subsessile; petiole sheath very short, ligulate, very thin, scarious at base. Leaf blade somewhat spongy, obovate-cuneate to obovate-oblong, apically rounded, truncate or retuse; midrib absent, primary veins subparallel, all arising from base, diverging towards apex, running into margin near apex, strongly prominent on lower surface, higher order venation reticulate. Inflorescence solitary, very small. Peduncle very short, pubescent. Spathe pubescent externally, glabrous internally, somewhat constricted centrally, lower margins united with each other and with ovary wall forming a tube, free margins between tube and limb folded between stigma and male organs forming a partition between an upper male chamber and a lower female one; limb erect, expanded, acute-acuminate. Spadix shorter than spathe, mostly adnate to spathe, only the apical male zone free; female zone with a single female flower at base and a thin, pouch-shaped flap just below the spathe partition; male zone subtended by a thin, marginally lobed, annular flap, composed of a basally naked spadix axis supporting a single whorl of 2–8 flowers, axis sometimes extending beyond. Flowers unisexual, perigone absent. Male flowers composed of 2 united stamens, thecae dehiscing by single apical slit. Female flower obliquely adnate to spadix axis, ovary ovoid, 1-locular, ovules c.25, orthotropous, placenta broad, apparently parietal, probably morphologically basal; style attenuate, bending inwards towards male flowers; stigma small, discoid-subcapitate. Berry thin-walled, utricular, several-seeded, ellipsoid, irregularly breaking up and decaying to release seeds. Seed barrel-shaped, ± subtruncate, excavated at apex and base, testa thick, reticulate-alveolate.

A monotypic pantropical genus with great variation depending on local environmental conditions.

Pistia stratiotes L., Sp. Pl.: 963 (1753). —Wild in Kirkia **2**: 13 (1961). —Mayo in F.T.E.A., Araceae: 66, fig.17 (1985). —Cook, Aq. Wetl. Pl. Sthn. Africa: 61 (2004). Type: India, Kerala; 'Kodda Pail', Rheede, Hort. Malab.: 11, t.32 (1692), lectotypified by Nicolson (Fl. Ceylon **6**: 100, 1988). FIGURE 12.1.**13**.

Floating aquatic herb. Leaves 3–10 cm or more long, up to 7.5 cm wide, spathulate-obconic, apex truncate, green, pubescent on both surfaces, but more densely below. Inflorescence solitary, hidden among leaf bases. Spathe up to 2 cm long, paler than leaves, externally pubescent, margins pilose, internally glabrous. Spadix short, mostly adnate to spathe, with single female

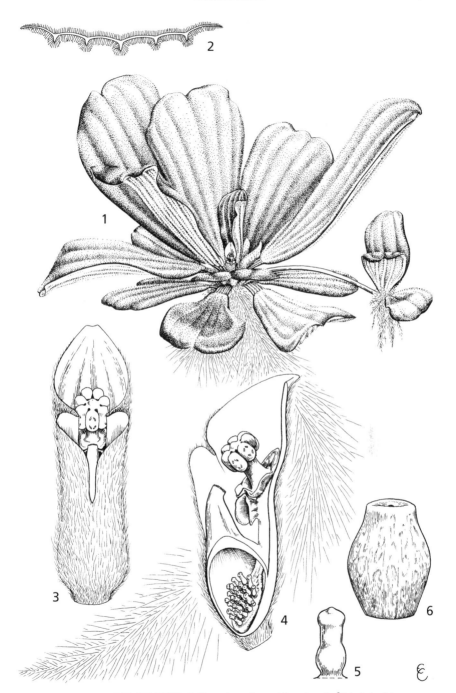

Fig. 12.1.**13**. PISTIA STRATIOTES. 1, flowering plant with stolon (× ²/₃); 2, leaf in transverse section (× 2); 3, inflorescence, front view (× 5); 4, inflorescence, side view with spathe and ovary wall part-removed (× 5); 5, ovule (× 32); 6, seed, side view (× 15). 1,2 from *Giles & Woolliams* PB30 (cult, K), 3–5 from *Wild* in *SRGH* 26411, 6 from *Greenway & Kanuri* 15152. Drawn by Eleanor Catherine. From F.T.E.A.

flower and 2–8 male flowers in a single whorl. Berry several seeded; seeds red-brown, barrel shaped, up to 2 mm long.

Botswana. N: Okavango, Motsaudi Is., fl. 20.iii.1973, *P.A. Smith* 470 (K). **Zambia**. N: Kaputa Dist., Mwawe dambo, N end of Lake Mweru-Wantipa, 17.iv.1961, *Phipps & Vesey-Fitzgerald* 3258 (K). E: Petauke Dist., 8 km W of Petauke, 30.iv.1961, *Leach & Rutherford* 11096 (K). S: Kalomo Dist., Kazungula, backwater of Zambezi R., 25.xi.1949, *Wild* 26411 (K). **Zimbabwe**. N: Chirundu Dist., Chirundu, iv.1948, *Whellan* 20036 (K, SRGH). W: Hwange Dist., Zambezi R., 10.iii.1976, *Gonde* 49 (K). C: Harare Dist., Lake Chivero, by yacht club, *M.A. Hyde* (sight record, Fl. Zimbabwe website). S: Chiredzi Dist., Gonarezou Nat. Park (Game Reserve), near Save/Runde river junction, 31.v.1971, *Ngoni* 142 (K). **Malawi**. N: Rumphi Dist., S Rukuru R., below Lake Kazuni, 17.x.1973, *Pawek* 7399 (K, MO, SRGH, UC). C: Kasungu Dist., Bua R., 28.iv.1967, *Salubeni* 698 (K, LISC). S: Lower Shire, Chiromo, junction of Shire and Ruo rivers, 19.vii.1958, *Seagrief* 3097 (K). **Mozambique**. N: Mandimba Dist., Lake Amaramba, Rio Lugenda, 13.x.1942, *Mendonça* 813 (LISC). Z: Mocuba Dist., Mocuba, 10.iii.1943, *Torre* 4905 (BR, LISC, WAG). T: Mágoè Dist., 10 km from Mphende (Mágoè Velho) to Cachomba, 28.ii.1970, *Torre & Correia* 18142 (LISC). MS: Cheringoma Dist., Inhamitanga, 20.ii.1948, *Andrada* 1078 (LISC). GI: Xai-Xai Dist., near Xai-Xai, Chibuto, 10.x.1958, *Mogg* 32625 (LISC).

Pantropical, also in South Africa. Open, still freshwater habitats; 20–1200 m.

Conservation notes: Introduced from South America; a pantropical aquatic weed. A serious aquatic weed in some places, often known as Water Lettuce; declared a pest in South Africa.

14. REMUSATIA Schott

Remusatia Schott in Schott & Endlicher, Melet. Bot.: 18 (1832). —Mayo, Bogner & Boyce, Gen. Araceae: 280 (1997).

Small to medium-sized, seasonally dormant herbs, tuber subglobose, producing erect to spreading, unbranched or branching stolons from axils of cataphylls, stolons producing small, ovoid tubercles at nodes, each with numerous, apically hooked scales. Leaves 1–2; petiole ± slender, sheath relatively short; blade peltate, cordate-lanceolate to cordate-ovate, acuminate; basal ribs well-developed, primary lateral veins pinnate, forming submarginal collective vein very close to margin, marginal vein also present, secondary and tertiary laterals arising from primaries at a wide angle, then arching towards leaf margin and forming inconspicuous interprimary collective veins, higher order venation reticulate. Inflorescence solitary, appearing with or without leaf. Peduncle shorter than petiole. Spathe strongly constricted between tube and blade, sometimes with secondary constriction above spadix; tube with convolute margins, persistent, enclosing female zone and sterile zone of spadix, blade longer than tube, fully expanded or remaining convolute and opening only at base, sometimes becoming reflexed, later deciduous. Spadix sessile, much shorter than spathe, female zone subcylindric, ± half as long as spathe tube, separated from male zone by much narrower zone of sterile male flowers, male zone ellipsoid or subclavate, fertile to apex, obtuse. Flowers unisexual, perigone absent. Male flowers 2–3-androus, stamens

Fig. 12.1.14. REMUSATIA VIVIPARA. 1, habit (× ^{1}/$_9$); 2, plant with inflorescence and bulbil-axes (× ½); 3, detail of bulbil cluster (× 2); 4, bulbil (× 4); 5, leaf (× ½); 6, detail of leaf showing marginal venation (× ^{2}/$_3$); 7, inflorescence (× ^{2}/$_3$); 8, spadix (× 1); 9, male flowers, side view (× 8); 10, male flower from above (× 8); 11, sterile male flowers from above (× 8); 12, female flowers, side view and longitudinal section (× 8); 13, sterile pistillodes at extreme base of spadix, side view (× 8). 1,2 from photos at Kew, 3 from *Brummitt et al.* 14015, 4 from *Deighton* 3824, 5,6 from plant cultivated at Kew 577-68, 7–13 from Kew spirit collection 19155. Drawn by Eleanor Catherine. From F.T.E.A.

connate into a cuneate-clavate, 4–6-grooved synandrium, fused filaments distinct, common connectives somewhat excavated at apex, thecae 4–6, oblong to ellipsoid, dehiscing by an apical pore-like slit; sterile male flowers each a ± elongated synandrode. Female flowers: staminodes absent, ovary broadly ovoid to subglobose, 1-locular or partially 2–4-locular at apex, ovules many, hemiorthotropous, funicle short to long, placenta 2–4 and parietal or 1 and basal; style very short or lacking; stigma discoid-subcapitate or slightly 3–4-lobed. Infructescence ellipsoid, borne within the persistent spathe tube; berry obovoid to globose, many-seeded. Seeds ellipsoid to subglobose.

A genus of 4 species, all native to the Himalayas; *R. vivipara* is widespread in the Old World tropics.

Remusatia vivipara (Roxb.) Schott in Schott & Endlicher, Melet. Bot.: 18 (1832). — Mayo in F.T.E.A., Araceae: 40 (1985). Type: India, Kerala, illustration in Rheede, Hort. Malab.: 12, t.9 (1693). FIGURE 12.1.**14**.

Arum viviparum Roxb., Hort. Bengal.: 65 (1814) as *viviparium*.

Tuber depressed globose, 2–3 × 1.5–2.5 cm. Stolons erect, simple, 14–22 × 0.5 cm, with tubercles on upper half. Cataphylls 7–13 cm long. Leaves 1(3); petiole cylindric, 30–35 cm long; blade 21–30 × 14–18 cm, peltate, ovate (insertion of petiole in lower third), cordate at base, cuspidate at apex, bright green, 4–6 primary lateral veins on each side, forming a 45° angle with midrib, posterior lobes up to half the length of anterior lobe, sinus rounded. Inflorescence rarely seen. Peduncle 6–15 cm long. Spathe 7–13.5; tube ovoid, 2–4.5 × 1 cm, green, strongly constricted at apex, limb subcircular to broadly spathulate, 6–9 × 5–6.5 cm, creamy yellow, reflexed at maturity. Spadix 3–4.5 cm long; female zone subcylindric, green, 1.8–2 × 0.8; intermediate sterile zone conic, c.1.1 × 0.2 cm; male zone clavate, 0.7–1.5 cm long. (Inflorescence description from Bogner, Fl. Madagas., Aracees: 28, 1975 and F.T.E.A.) Berries ovoid, enclosed by spathe; seeds numerous, elliptic.

Zambia. W: Mwinilunga Dist., West Lunga R., 8 km N of Mwinilunga, st. 23.i.1975, *Brummitt, Chisumpa & Polhill* 14015 (K); Luakera Falls, 25.i.1938, *Milne-Redhead* 4329 (K).

Found in S Asia, China and Australia, Madagascar, West and Central Africa, Ethiopia and Tanzania. Rock faces in riverine forest; 1200–1400 m.

Conservation notes: In the Flora area known only from NW Zambia although the global distribution is very widespread; not threatened.

No flowering material from Africa has been seen, but this may be due to undercollection rather than rarity of flowering. The tubercles seem to be an effective means of dispersal as the species has a very wide range at higher altitudes in the Old World tropics.

15. **SAUROMATUM** Schott

Sauromatum Schott in Schott & Endlicher, Melet. Bot.: 17 (1832). —Mayo, Bogner & Boyce, Gen. Araceae: 263 (1997).

Small to medium-sized geophytic herbs, rarely pubescent. Tuber subglobose or depressed-globose, sometimes large. Leaves solitary to several. Petiole occasionally spotted or marked, sometimes pilose, sheath very short. Leaf blade very variable, ovate, cordate to hastate, more usually pedatifid to pedatisect; primary lateral veins of lobes pinnate, forming a submarginal collective vein, 1–2 marginal veins also present, higher order venation reticulate. Inflorescence solitary, appearing with

Fig. 12.1.**15**. SAUROMATUM VENOSUM. 1, tuber (× ⅙); 2, plant in leaf (× ⅙); 3, central leaf lobe (× ⅔); 4, flowering plant (× ⅙); 5, lower part of spadix, spathe partly removed (× 1); 6, stamen, side view (× 10); 7, stamen from above (× 10); 8, sterile projections (× 3); 9, female flower, side view (× 10); 10, female flower, longitudinal section showing ovules and placentation (× 10); 11, female flower, from above (× 10); 12, infructescence (× ⅓); 13, berry, side view (× 2), 14, berry, longitudinal section showing seed (seed-coat hatched) (× 2). 2–11 from *Milne-Readhead & Taylor* 10446, 12–14 from *Tanner* 692. Drawn by Ann Webster. From F.T.E.A.

or without leaves; peduncle shorter than petiole. Spathe constricted between tube and limb; tube with convolute or united margins, ± cylindric; limb longer, broadly ovate or triangular to lanceolate, erect to reflexed and spiralled-revolute, rarely margins undulate. Spadix sessile, shorter than or subequal to spathe, free; female zone conic to cylindric, separated from male zone by longer axis with either sterile flowers in lower part and naked above or entirely covered in sterile flowers; male zone cylindric, densely flowered; appendix stipitate or not, slender and elongate to stout and short-conic. Flowers unisexual, perigone absent; male flowers few-androus, stamens free, anthers ± sessile, connective slender or rarely broad, thecae variously shaped; sterile flowers either all similar or diverse on the same spadix, variously shaped, usually at least some clavate; female flowers with gynoecium globose to cylindric, ovary 1-locular, ovules 1–4, orthotropous, funicle short, placenta basal, sessile to subsessile, stigma depressed to discoid-hemispheric. Berries densely congested in subglobose, sometimes partly hypogeal infructescence. Seeds globose to ovoid, testa thin, smooth to rough.

A genus of about 9 species from China and SE Asia to tropical Africa, with most diversity in China (Yunnan).

Although previously included within *Typhonium*, *Sauromatum* is now considered to be a separate genus on the basis of recent DNA analyses (Cusimano *et al.* in Taxon **59**: 439–447, 2010).

Sauromatum venosum (Aiton) Kunth, Enum. Pl. **3**: 28 (1841). —Hepper in F.W.T.A., ed.2 **3**: 116 (1968). —Mayo in F.T.E.A., Araceae: 58 (1985). Type: Unknown origin, plant cultivated at Kew by W. Malcolm in 1774 (BM† holotype). FIGURE 12.1.**15**.

Arum venosum Aiton in Hort. Kew **3**: 315 (1789).
Typhonium venosum (Aiton) Hett. & P.C. Boyce in Aroideana **23**: 51 (2000).

Seasonally dormant herb. Tuber depressed-globose, 4–10 cm in diameter. Leaf solitary; petiole smooth, up to 90 cm long, green, often black-purple spotted or tinged brownish purple; blade kidney-shaped to subcircular in outline, 25–55 cm broad, pedatifid; lobes 7–11, ovate-elliptic to ovate-lanceolate, acuminate, cuneate, deeply divided almost to base, central lobe usually obovate-elliptic, 16–40 × 7–15 cm. Inflorescence appearing before leaves, with very strong carrion smell at anthesis, subtended by several oblong-lanceolate cataphylls; peduncle short, 2–8 cm when in flower, subterranean, spathe borne at or partly below ground-level. Spathe 28–45 cm long; tube with joined margins, subcylindric with inflated ellipsoid basal part, 4–11 × 2–5 cm, outside dull purplish green, inside reddish purple, smooth; limb narrowly lanceolate, 22–35 × 3–6 cm, reflexed and twisting spirally at maturity, inner surface pale yellow to pale green with numerous dark maroon-purple markings, margins undulating. Spadix sessile, 19–43 cm long; female zone subcylindric, 1–1.8 × 0.6–1 cm; sterile zone with cylindric longitudinally ridged axis, 3.2–6.5 × 0.3–0.5 cm, cream becoming purple apically, with 0.3–0.8 cm long clavate, sterile projections at base; male zone cylindric, 0.9–1.4 × 0.5–0.9 cm, sulphur yellow; sterile appendix cylindric, 12–30 × 0.5–1.2 cm at base, tapering upwards, yellow to greenish brown, erect, curving forward at anthesis. Female flowers: ovary flask-shaped to cylindric, dark reddish-purple; ovules 1–2, pear-shaped, on massive basal placenta; stigma capitate, cream, 0.3–0.5 mm in diameter. Infructescence subglobose, at or partly below ground level, up to 4 cm wide. Berries obovoid, 0.9–1 × c.0.8 cm, 1–2 seeded, deep purple. Seeds obovate, 0.7–0.8 × 0.4–0.5 cm; testa fleshy, dark spotted.

Zambia. N: Mbala Dist., Ndundu riverine forest, st. 4.v.1962, *Richards* 16439 (K). W: Solwezi, fl. 13.ix.1952, *Angus* 441 (K). **Malawi.** N: Chitipa Dist., Misuku Hills, Chisasu-Itera road, st. 27.xii.1977, *Pawek* 13393 (K, MO). C: Dedza Dist., near Bembeke Mission, fl. 15.xi.1967, *Salubeni* 896 (K, SRGH). S: Blantyre Dist., Bangwe Hill, 4 km E of Limbe, st. 23.xi.1977, *Brummitt et al.* 15157 (K). **Mozambique.** Z: Gúruè Dist., Serra do Gúruè, fl. 19.ix.1944, *Mendonça* 2107 (LISC).

Also in South-central Asia, Arabian Peninsula, W Africa from Cameroon to Central African Republic, Ethiopia, Uganda, Kenya, Tanzania, Congo and Angola. In shade, in riverine and evergreen forest, on humus rich soil; 900–1900 m.

Conservation notes: Widespread species; not threatened.

INDEX TO BOTANICAL NAMES

FAMILIES OF VASCULAR PLANTS REPRESENTED IN THE FLORA ZAMBESIACA AREA

PTERIDOPHYTA
(Flora Zambesiaca families and family number. Published 1970)

Actiniopteridaceae		Gleicheniaceae	9	Parkeriaceae			
see Adiantaceae	18	Grammitidaceae	20	see Adiantaceae	18		
Adiantaceae	18	Hymenophyllaceae	15	Polypodiaceae	21		
Aspidiaceae	27	Isoetaceae	4	Psilotaceae	1		
Aspleniaceae	23	Lindsaeaceae	19	Pteridaceae			
Athyriaceae	25	Lomariopsidaceae	26	see Adiantaceae	18		
Azollaceae	13	Lycopodiaceae	2	Salviniaceae	12		
Blechnaceae	28	Marattiaceae	7	Schizaeaceae	10		
Cyatheaceae	14	Marsileaceae	11	Selaginellaceae	3		
Davalliaceae	22	Oleandraceae		Thelypteridaceae	24		
Dennstaedtiaceae	16	see Davalliaceae	22	Vittariaceae	17		
Dryopteridaceae		Ophioglossaceae	6	Woodsiaceae			
see Aspidiaceae	27	Osmundaceae	8	see Athyriaceae	25		
Equisetaceae	5						

GYMNOSPERMAE
(Flora Zambesiaca families and family number. Volume 1(1) 1960)

Cupressaceae	3	Cycadaceae	1	Podocarpaceae	2

ANGIOSPERMAE
(Flora Zambesiaca families, volume and part number and year of publication)

Acanthaceae	-	-	Balsaminaceae	2(1)	1963
Agapanthaceae	13(1)	2008	Barringtoniaceae	4	1978
Agavaceae	13(1)	2008	Basellaceae	9(1)	1988
Aizoaceae	4	1978	Begoniaceae	4	1978
Alangiaceae	4	1978	Behniaceae	13(1)	2008
Alismataceae	12(2)	2009	Berberidaceae	1(1)	1960
Alliaceae	13(1)	2008	Bignoniaceae	8(3)	1988
Aloaceae	12(3)	2001	Bixaceae	1(1)	1960
Amaranthaceae	9(1)	1988	Bombacaceae	1(2)	1961
Amaryllidaceae	13(1)	2008	Boraginaceae	7(4)	1990
Anacardiaceae	2(2)	1966	Brexiaceae	4	1978
Anisophylleaceae			Bromeliaceae	13(2)	2010
see Rhizophoraceae	4	1978	Buddlejaceae		
Annonaceae	1(1)	1960	see Loganiaceae	7(1)	1983
Anthericaceae	13(1)	2008	Burmanniaceae	12(2)	2009
Apocynaceae	7(2)	1985	Burseraceae	2(1)	1963
Aponogetonaceae	12(2)	2009	Buxaceae	9(3)	2006
Aquifoliaceae	2(2)	1966	Cabombaceae	1(1)	1960
Araceae	12(1)	2012	Cactaceae	4	1978
Araliaceae	4	1978	Caesalpinioideae		
Aristolochiaceae	9(2)	1997	see Leguminosae	3(2)	2006
Asclepiadaceae	-	-	Campanulaceae	7(1)	1983
Asparagaceae	13(1)	2008	Canellaceae	7(4)	1990
Asphodelaceae	12(3)	2001	Cannabaceae	9(6)	1991
Avicenniaceae	8(7)	2005	Cannaceae	13(4)	2010
Balanitaceae	2(1)	1963	Capparaceae	1(1)	1960
Balanophoraceae	9(3)	2006	Caricaceae	4	1978

Molluginaceae	4	1978		Rhizophoraceae	4	1978
Monimiaceae	9(2)	1997		Rosaceae	4	1978
Montiniaceae	4	1978		Rubiaceae		
Moraceae	9(6)	1991		subfam. Rubioideae	5(1)	1989
Musaceae	13(4)	2010		tribe Vanguerieae	5(2)	1998
Myristicaceae	9(2)	1997		subfam.Cinchonoideae	5(3)	2003
Myricaceae	9(3)	2006		Rutaceae	2(1)	1963
Myrothamnaceae	4	1978		Salicaceae	9(6)	1991
Myrsinaceae	7(1)	1983		Salvadoraceae	7(1)	1983
Myrtaceae	4	1978		Santalaceae	9(3)	2006
Najadaceae	12(2)	2009		Sapindaceae	2(2)	1966
Nesogenaceae	8(7)	2005		Sapotaceae	7(1)	1983
Nyctaginaceae	9(1)	1988		Scrophulariaceae	8(2)	1990
Nymphaeaceae	1(1)	1960		Selaginaceae		
Ochnaceae	2(1)	1963		see Scrophulariaceae	8(2)	1990
Olacaceae	2(1)	1963		Simaroubaceae	2(1)	1963
Oleaceae	7(1)	1983		Smilacaceae	12(2)	2009
Oliniaceae	4	1978		Solanaceae	8(4)	2005
Onagraceae	4	1978		Sonneratiaceae	4	1978
Opiliaceae	2(1)	1963		Sphenocleaceae	7(1)	1983
Orchidaceae	11(1)	1995		Sterculiaceae	1(2)	1961
Orchidaceae	11(2)	1998		Strelitziaceae	13(4)	2010
Orobanchaceae				Taccaceae		
see Scrophulariaceae	8(2)	1990		see Dioscoreaceae	12(2)	2009
Oxalidaceae	2(1)	1963		Tecophilaeaceae	12(3)	2001
Palmae	13(2)	2010		Tetragoniaceae	4	1978
Pandanaceae	12(2)	2009		Theaceae	1(2)	1961
Papaveraceae	1(1)	1960		Thymelaeaceae	9(3)	2006
Papilionoideae				Tiliaceae	2(1)	1963
see Leguminosae	-	-		Trapaceae	4	1978
Passifloraceae	4	1978		Turneraceae	4	1978
Pedaliaceae	8(3)	1988		Typhaceae	13(4)	2010
Periplocaceae				Ulmaceae	9(6)	1991
see Asclepiadaceae	-	-		Umbelliferae	4	1978
Philesiaceae				Urticaceae	9(6)	1991
see Behniaceae	13(1)	2008		Vacciniaceae		
Phormiaceae				see Ericaceae	7(1)	1983
see Hemerocallidaceae	12(3)	2001		Vahliaceae	4	1978
Phytolaccaceae	9(1)	1988		Valerianaceae	7(1)	1983
Piperaceae	9(2)	1997		Velloziaceae	12(2)	2009
Pittosporaceae	1(1)	1960		Verbenaceae	8(7)	2005
Plantaginaceae	9(1)	1988		Violaceae	1(1)	1960
Plumbaginaceae	7(1)	1983		Viscaceae	9(3)	2006
Podostemaceae	9(2)	1997		Vitaceae	2(2)	1966
Polygalaceae	1(1)	1960		Xyridaceae	13(4)	2010
Polygonaceae	9(3)	2006		Zannichelliaceae	12(2)	2009
Pontederiaceae	13(2)	2010		Zingiberaceae	13(4)	2010
Portulacaceae	1(2)	1961		Zosteraceae	12(2)	2009
Potamogetonaceae	12(2)	2009		Zygophyllaceae	2(1)	1963
Primulaceae	7(1)	1983				
Proteaceae	9(3)	2006				
Ptaeroxylaceae	2(2)	1966				
Rafflesiaceae	9(2)	1997				
Ranunculaceae	1(1)	1960				
Resedaceae	1(1)	1960				
Restionaceae	13(4)	2010				
Rhamnaceae	2(2)	1966				